One theorem started it all

Jesse Sakari Hyttinen

# One theorem started it all

From treespeak to famous numbers

Publisher: BoD – Books on Demand, Helsinki, Finland

Producer: BoD – Books on Demand, Norderstedt, Germany

ISBN: 978-952-80-4693-6

For my father, mother, big sister, Elle and Eppu

# Contents

0.Introduction......................................................13


1.The secret language of trees...................................14

1.1Theorem of sum forms........................................16

1.2Tget.........................................................20

1.2.1Condition matrix.............................................20

1.2.2Vertex edge algebra........................................23

1.2.2.1Symbols....................................................23

1.2.2.2Rules......................................................38

1.2.2.3Known graphical equalities.............................44

1.2.3Tree generation algorithm and its variations....48

1.2.3.1Tree generation algorithm.............................49

1.2.3.2Examples.................................................55

1.2.3.3Information about the tree generation
algorithm.........................................................72

1.2.3.4Branch form generator and tree enumeration algorithm…………………………………………….74

1.2.3.5Branch form generator…… …………………………..74

1.2.3.6Tree enumeration algorithm……………………….75

1.2.3.7Examples……………………………………………….76

1.2.3.8Enumeration formula for rooted trees…………101

1.3Treespeak……………………………………………….104


2.Controversial math……………………………………….106

2.1The odd zero of a modified Riemann zeta function……………………………………………………….106

2.2Series deduction method and dimension theory………………………………………………………111

2.2.1Dimension theory……………………………………111

2.3Derivative of a discrete function ……………………131

3.Good old constant.................................................137

3.1Pi.....................................................................137

3.2Euler's number e...............................................140

3.3Euler-Mascheroni constant.................................142


4.A theorem in geometry .......................................144


5.Identity number and the golden ratio...................151


6.A theorem for number one...................................165

# 0.Introduction

A repeating theme among my own ideas is a mathematical tree, the focus of my work. During the years, the ideas have accumulated and culminated into this work – the most important one so far. I hope that this book will inspire you and awake your curiosity, as mathematics is full of opportunities – one just must have perseverance and a will to succeed! Knowledge and skills will increase when one works hard and retries again and again. If I would have quit working on these ideas during my main research years due to a failure, then you would not be reading this book right now. Failure may really be an almost necessary condition to success. The key factor is that you learn from your mistakes, and thus avoid repeating the same ones once again.

If mathematics is your thing, then nothing should be stopping you from achieving your dreams! If I have developed these ideas to the stage they are now at without much support from anyone, then what is your excuse? Believe in yourself, as I did in myself and still do.

# 1.The secret language of trees

In the year 2014 I started to research a one particular problem in graph theory. The problem was from the film Good Will Hunting:

Draw all homeomorphically irreducible trees of size n = 10.

So, one had to draw all the size 10 series reduced free trees. May sound complicated, but it actually is quite simple. The following picture has the solution:

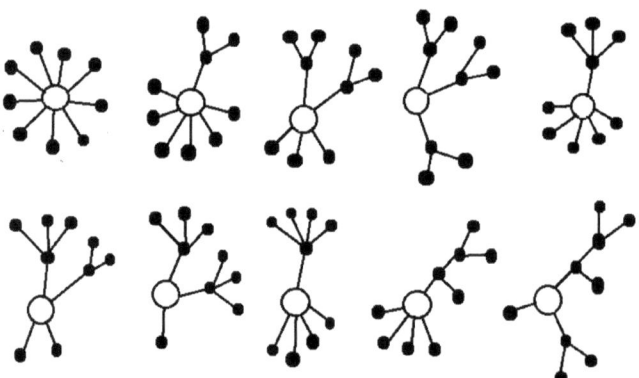

See those objects that consist of white and black vertices and the edges which connect the vertices to each other? They are trees, series reduced free trees in this case. If you count the vertices in each tree, you will notice that every one of them has ten each. In other words, size 10 trees. The white vertices may seem a bit odd among the many black ones, but they do have a purpose: They represent the so-called root. A virtual root in this case, as free trees have no real roots in mathematics.

The root / virtual root is used in my ideas as a base of a tree when vertex edge algebra – the language of mathematical trees I have created – constructs the said tree. In the year 2020 I finally created the theorem that had already been present in my theories since 2014. Thanks to this theorem, vertex edge algebra can be used as a tree language. I call the theorem the theorem of sum forms, as vertex edge algebra constructs trees as partitions of positive, nonzero integers: sums of numbers, that now represent forms of trees.

# 1.1Theorem of sum forms

" Any nonempty, finite tree can be expressed with a combination consisting of three object types: ones 1, plus-operators + and brackets ( )."

## A sort of a proof:

Any nonempty, finite tree consists of n vertices and $n - 1$ edges. The tree is connected so there are no isolated vertices (except for a single vertex, size one tree) and between any two arbitrarily chosen vertices there exists exactly one, unique path. Thus, we can choose a root/ virtual root, which is connected to every other vertex in the tree via a path. Let us express the root and every other vertex with a number one, edges with plus-operators and a starting branch with brackets in a way that the root always starts the tree, and an internal vertex starts every branch. External vertices are the last objects inside / outside branches and consecutive external vertices at the same level of depth are connected to the root (outside brackets) / internal vertex (inside brackets), which starts that level of depth.

We construct a sum form, in which the root is connected via a path consisting of consecutive edges (+ operators) to any vertex in the tree. Examples:

1

Single root, size one tree.

$1 + 1 + 1$

Two external vertices are connected to the root. Size three tree.

$1 + (1 + 1)$

The root is connected to an internal vertex (the first number one inside the brackets), and this internal vertex is connected to an external vertex (the last number one inside the brackets). Size three tree.

$1 + (1 + 1) + 1 + 1$

The same as in the previous example, but the root is additionally connected to two new external vertices. Size five tree.

$$1 + (1 + (1 + 1) + 1) + (1 + 1) + 1$$

In this case one should check out the picture, as writing and understanding the description may get quite laborious. Size eight tree.

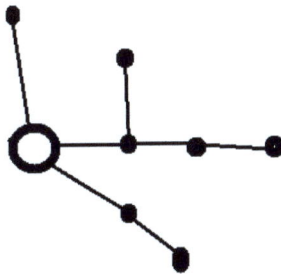

Now as the theorem of sum forms has been presented, the collection of three main ideas I have created can now be examined. Vertex edge algebra is one of these ideas.

# 1.2Tget

Tget, tree generation and enumeration triplet, is a collection of three main ideas, which are the condition matrix, vertex edge algebra and tree generation algorithm. The algorithm also has two additional varieties: Branch form generator and tree enumeration algorithm.

## 1.2.1Condition matrix

Condition matrix $C_{v+x_i} = \{c_k\}_k$, is a collection of conditions $c_k$ ($k \in \mathbb{N}$), where v is the sum of a tree's vertices (the types counted are defined by the condition v'), in a forest defined by a type $x_i$ condition matrix. The conditions together usually define the type of trees the forest has; the idea would be that only with knowing the properties of a tree could one find its forest and name in a possible database of condition matrices.

Tget focuses on four tree types:

1)Rooted trees

$$C_{n+x_{0.0}} = \{r_v[+], v = n, v' = rie, d_r \geq 1, d_i \geq 2, d_e = 1\} \quad (1)$$

2)Series-reduced rooted trees

$$C_{n+x_{0.1}} = \{r_v[+], v = n, v' = rie, d_r \neq 2, d_i \geq 3, d_e = 1\} \quad (2)$$

3)Free trees

$$C_{n+x_{1.0}} = \{r_v[-], v = n, v' = rie, d_r \geq 1, d_i \geq 2, d_e = 1\} \quad (3)$$

4)Series-reduced free trees

$$C_{n+x_{1.1}} = \{r_v[-], v = n, v' = rie, d_r \neq 2, d_i \geq 3, d_e = 1\} \quad (4)$$

$r_v[+]$ means that the root is treated as a root, so it is not a virtual root in this case, but a vertex with a label. $r_v[-]$, on the other hand, means that the root is virtual and only acts as a base for vertex edge algebraic operations. The same acting as a base also goes for the case $r_v[+]$, in addition to being as a root.

v is the sum of the vertices and v′ defines the types of vertices that are counted to create v – in this case, v′ = rie (root, internal, external vertex), so the vertex types which are counted are in this case the root, internal and external vertices. Thus, the number v is in this case the size of a tree in the forest. $d_r$ is the number of vertices which are linked to the root, in other words it is the degree of the root. $d_i$ is, on the other hand, the degree of the internal vertex and $d_e$ is the degree of the external vertex. These have naturally positive nonzero integer values, except for the single root, size one tree where $d_r = 0$.

## 1.2.2Vertex edge algebra

Trees can be expressed as sums of numbers, allowing the construction of a mathematical language called vertex edge algebra. This language is used by the tree generation algorithm and its varieties – branch form generator and tree enumeration algorithm. The symbols and rules for this language will now be presented.

## 1.2.2.1Symbols

$1^o$ is the root, which always starts the vertex edge algebraic representation of a tree – the sum form. There is exactly one root per one tree. For further clarification, one can make a notion $1^{o+}$, if the sum form obeys the condition $r_v[+]$, and a notion $1^{o-}$ , if the sum form obeys the condition $r_v[-]$. If the condition is $r_v[+]$, then no two different sum forms have the same graphical representation, due to the real root. But, if the condition is $r_v[-]$, then the virtual root makes it possible that two different sum forms may have the same graphical representation.

$1^*$ is an internal vertex and always starts a branch by being the first object inside the brackets. There is always one internal vertex per one bracket pair, although there can be several internal brackets inside brackets and thus several internal vertices, for example:

$$1^o + \left(1^* + \left(1^* + (1^* + 1^-) + (1^* + 1^- + 1^-)\right)\right) + 1^- .$$

$1^-$ is an external vertex, which is always the last object at every depth level, when every new internal bracket pair starts a new level.

$+$ is an edge and is always between two branches, two vertices, a vertex and a branch or a branch and a vertex. Outside brackets a branch or a vertex is connected with an edge to the root, whereas inside the brackets to the internal vertex which starts those brackets.

( ) are brackets, which always start a branch that has a size greater than one (two or more vertices). The first object inside the brackets is always an internal vertex.

· is a multiplication sign, which can make sum forms more compact. The notion $n^{\sim} \cdot m$ means, that there are $m$ branches $n^{\sim}$ and $m - 1$ concealed $+$ operators. For example,

$$1^o + 3^x \cdot 2 + 1^- \cdot 3 = 1^o + 3^x + 3^x + 1^- + 1^- + 1^-.$$

$n^x$ $(n \geq 3)$ is a branch, in which there is an internal vertex connected to $n - 1$ external vertices. In other words,

$$n^x = \left(1^* + 1^- \cdot (n - 1)\right).$$

Some examples are $3^x = (1^* + 1^- \cdot 2)$ and $4^x = (1^* + 1^- \cdot 3)$.

$n^-$ ($n \geq 2$), on the other hand, is a branch where there are $n - 1$ consecutive internal vertices connected to each other in a line and one external vertex in the end. In other words,

$$n^- = \left(1^* + \left(1^* + (1^* + \ ... + (1^* + 1^-)\ ...)\right)\right)$$

, in which there are $n - 1$ bracket pairs and $n - 1$ internal vertices. Some examples are $2^- = (1^* + 1^-)$ and $3^- = (1^* + (1^* + 1^-))$.

The sum form in the picture: $1^o$

The sum form in the picture: $1^0 + 1^-$

The sum form in the picture: $1^0 + (1^* + 1^-) = 1^0 + 2^-$

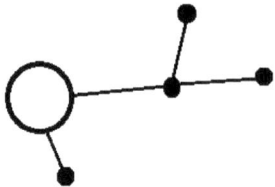

The sum form in the picture: $1^o + (1^* + 1^- \cdot 2) + 1^- = 1^o + 3^x + 1^-$

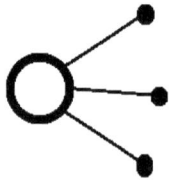

The sum form in the picture: $1^o + 1^- \cdot 3$

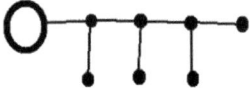

The sum form in the picture: $1^o + (1^* + (1^* + 3^x + 1^-) + 1^-)$.

$n^{mL}$ ($n \geq 3, n \geq 2m + 1, m \geq 0$) is a branch, in which there is an internal vertex connected to m branches $2^-$ and also $n - 2m - 1$ external vertices. In other words,

$$n^{mL} = (1^* + 2^- \cdot m + 1^- \cdot (n - 2m - 1))$$

$$(2m + 1)^{mL} = (1^* + 2^- \cdot m)$$

$$n^{1L} = n^L$$

$$n^{0L} = n^x$$

$$3^{1L} = 3^-.$$

Some examples are $3^{1L} = (1^* + 2^-) = 3^-$ , $4^{1L} = (1^* + 2^- + 1^-)$, $5^{2L} = (1^* + 2^- \cdot 2)$, $7^{2L} = (1^* + 2^- \cdot 2 + 1^- \cdot 2)$, $7^{0L} = 7^x$.

$n^Y (n \geq 4)$ is a branch, in which there are $n - 3$ consecutive internal vertices connected to each other in a line, and a branch $3^x$ in the end. In other words,

$$n^Y = \left(1^* + \left(1^* + (1^* + ... + (1^* + 3^x) ...)\right)\right)$$

, in which there are $n - 3$ bracket pairs and internal vertices. Some examples are $4^Y = (1^* + 3^x)$, $5^Y = (1^* + (1^* + 3^x))$, $6^Y = \left(1^* + \left(1^* + (1^* + 3^x)\right)\right)$.

$\doteq$ is a vertex sum equivalence notation and means that the sum forms on the both sides of the notation have the same vertex sums v (v' defines what types are counted to create v).

For example, when v' = rie,

$$1^o + 1^- \cdot 2 \doteq 1^o + 2^-$$

, as $1 + 1 \cdot 2 = 1 + 2 = 3$.

It is good to know, that even though the vertex sums are the same, these sum forms are otherwise different (except for possible graphical equivalence). Generally speaking, a sum form is both identical and equal in vertex sum only with itself. In vertex edge algebra, when using the algorithms, one generates and enumerates all the different sum forms, and any identical ones are not allowed.

$\equiv$ is a forest equivalence notation, which is used between a forest's closed and opened forms. For example, for a rooted tree size 4 forest,

$$\{4\}_{rie+}^{\Sigma o} \equiv 1^o + 1^- \cdot 3 \doteq 1^o + 2^- + 1^- \doteq 1^o + 3^x \quad (5)$$
$$\doteq 1^o + 3^-$$

, where $\{4\}_{rie+}^{\Sigma o}$ is the forest's closed form (with conditions v′ = rie and $r_v[+]$) and the right side of the notation $\equiv$ is the opened form of the forest.

$\cong$ is the graphical representation equivalence notation and is used between two sum forms, which have the same graphical representations. In the case $r_v[+]$ the sum form is graphically equivalent only with itself, the case $r_v[-]$ may be different. For example,

$$1^{o-} + 1^- \cdot 3 \cong 1^{o-} + 3^x \text{ and}$$

$$1^{o-} + 2^- \cdot 2 \cong 1^{o-} + 4^- \cong 1^{o-} + 3^- + 1^-.$$

$0^{()}$ is an empty branch, which has no real graphical representation. The size of an empty branch is zero, whereas the size of a branch $n^{\sim}$ is n.

$n^{a.-b.}$ is a notation, which is used in enumeration. It makes enumeration efficient and means, that the forms a. – b. of the branch $n$˜are gone through. For example, the case (5) can be rewritten as the following:

$$\{4\}_{rie+}^{\Sigma o} \equiv 1^o + 1^- \cdot 3 \doteq 1^o + 2^- + 1^- \quad (6)$$
$$\doteq 1^o + 3^{1.-2.} \ .$$

In rooted trees and free trees case, if there is a branch $n$˜ ($n \geq 3$) with m forms, then the first form is $n^x$ and the last form is $n^-$:

$$n^{1.} = n^x$$

and

$$n^{m.} = n^- \ .$$

If $m \geq 4$, then additionally

$$n^{2.} = n^L ,$$

$$n^{p.} = n^{(p-1)L} \ (n \geq 3, n \geq 2p - 1, p \geq 1)$$

and

$$n^{(m-1).} = n^Y .$$

$n^\sim(a, b)$ is also a notation, which is used in enumeration. It means, that the quantity factors a − b of the branch $n^\sim$ are gone through. The branch $n^\sim$ should only have one form, to simplify the case and ease the enumeration. Also only the first branch which has only one form should be used in this notation, to further simplify the case. For example, in the series-reduced rooted and series-reduced free trees case, only the branch $3^x$ should be used with this notation, and the branch $4^x$ should be excluded.

For the rooted trees case v = 5,

$$\{5\}_{rie+}^{\Sigma o} \equiv 1^o + 1^- \cdot 4 \doteq 1^o + 2^- + 1^- \cdot 2$$

$$\doteq 1^o + 2^- \cdot 2$$

$$\doteq 1^o + 3^x + 1^-$$

$$\doteq 1^o + 3^- + 1^-$$

$$\doteq 1^o + 4^x$$

$$\doteq 1^o + (1^* + 2^- + 1^-)$$

$$\doteq 1^o + (1^* + 3^x)$$

$$\doteq 1^o + 4^- .$$

Or using the further notations as well:

$$\{5\}_{rie+}^{\Sigma o} \equiv 1^o + 1^- \cdot 4 \doteq 1^o + 2^- + 1^- \cdot 2$$

$$\doteq 1^o + 2^- \cdot 2$$

$$\doteq 1^o + 3^x + 1^-$$

$$\doteq 1^o + 3^- + 1^-$$

$$\doteq 1^o + 4^x$$

$$\doteq 1^o + 4^L$$

$$\doteq 1^o + 4^Y$$

$$\doteq 1^o + 4^- \; .$$

This can be simplified:

$$\{5\}_{rie+}^{\Sigma o} \equiv 1^o + 2^-(0,2) + 1^-(4,0) \quad (7)$$

$$\doteq 1^o + 3^{1.-2.} + 1^-$$

$$\doteq 1^o + 4^{1.-4.} \; .$$

The efficiency while using the enumeration tools grows quickly when the index v (in this case the tree size) and thus the rooted tree quantity $r_v$ grow in size. This goes also for the quantity of series-reduced rooted trees $s_v$, quantity of free trees $t_v$ and quantity of series-reduced free trees (homeomorphically irreducible trees) $h_v$.

Z is a number, which is used to express the sum form quantity for a certain row in enumeration. For example for the case (7):

$$\{5\}_{rie+}^{\Sigma o} \equiv 1^o + 2^-(0,2) + 1^-(4,0) \ (Z = 3)$$

$$\doteq 1^o + 3^{1.-2.} + 1^-(Z = 2)$$
$$\doteq 1^o + 4^{1.-4.}(Z = 4)$$

, so $r_5 = 3 + 2 + 4 = 9$.

# 1.2.2.2 Rules

Let there be trees $T_1$, $T_2$ and $T_3$ with sum forms $(v_1)^o_{rie}$, $(v_2)^o_{rie}$ and $(v_3)^o_{rie}$.

Thus

1) $T_i = T_i$, $(v_i)^o_{rie} \doteq (v_i)^o_{rie}$ , $(v_i)^o_{rie} \cong (v_i)^o_{rie}$ , $i = 1,2,3$.

2) If $T_1 = T_2$, then these trees are the same.

3) If $(v_1)^o_{rie} \doteq (v_2)^o_{rie}$ , then $v_1 = v_2$.

4) If $(v_1)^o_{rie} \cong (v_2)^o_{rie}$ , then the graphical representations of the trees are the same.

5) If $T_1 = T_2$ and $T_2 = T_3$, then $T_1 = T_3$.

6) If $(v_1)^o_{rie} \doteq (v_2)^o_{rie}$ and $(v_2)^o_{rie} \doteq (v_3)^o_{rie}$ , then $(v_1)^o_{rie} \doteq (v_3)^o_{rie}$.

7) If $(v_1)^o_{rie} \cong (v_2)^o_{rie}$ and $(v_2)^o_{rie} \cong (v_3)^o_{rie}$ , then $(v_1)^o_{rie} \cong (v_3)^o_{rie}$.

8) If a branch / external vertex has a quantity factor in a sum form, this quantity factor can come before or after the branch / external vertex. However, there must also be a multiplication sign between these both.

Examples:

$$1^o + 3 \cdot 3^x = 1^o + 3^x \cdot 3$$

$$1^o + (1 + 2 + 3) \cdot 6^- = 1^o + 6^- \cdot 1 + 6^- \cdot 2 + 6^- \cdot 3$$

$$1^o + (1^* + 3^x) \cdot 2 = 1^o + 2 \cdot (1^* + 3^x).$$

9) If there are two or more branches connected to the root or an internal vertex, then the arrangement of these (what comes first, what is in the middle and what comes last in the sum form etc.) in that specific vertex does not matter. However, the root always starts the sum form , an internal vertex always starts a branch (size larger than one) with brackets and an external vertex is always the last object inside and outside brackets.

Examples:

$$1^o + 2^- + 3^- + 1^- = 1^o + 3^- + 2^- + 1^-$$

$$1^o + (1^* + 4^x + 2^- \cdot 3) + 3^x = 1^o + 3^x + (1^* + 3 \cdot 2^- + 4^x).$$

10)     There is always one root per one sum form and one internal vertex per one bracket pair in a way that the internal vertex starts those brackets. However, there can be several brackets and internal brackets and thus several internal vertices at depth levels that are getting deeper and deeper.

11)     + operator, an edge, is only used between two branches, two vertices, a branch and a vertex or a vertex and a branch. Outside brackets the + operator connects the object on its right side to the root, which starts the sum form; inside the brackets the + operator connects the object on its right side to the internal vertex, which starts those brackets.

12)　　· operator is only used between a quantity factor and a branch, a branch and a quantity factor, a quantity factor and an external vertex or an external vertex and a quantity factor.

13)　　For rooted trees and free trees case, if there is a branch $n^{\sim}$ with $m \geq 2$ forms, then the form $n^{x}$ is the first and the form $n^{-}$ is the last form.

14)　　If an empty branch is connected to the root or an internal vertex, the case is treated as if no branch was connected to them in the first place. Even if there is a quantity factor or not.

Examples:

$$1^{o} + 2^{-} + 3^{x} + 0^{()} = 1^{o} + 2^{-} + 3^{x}$$

$$1^{o} + 3^{L} \cdot 5 + 0^{()} \cdot 4 = 1^{o} + 3^{L} \cdot 5 \,.$$

15)    If a branch or an external vertex has a quantity factor of one, then this quantity factor does not have to be shown. If, however, the branch or external vertex has a quantity factor of zero, then the branch or external vertex does not exist in the sum form. In this case, neither the branch / external vertex nor the quantity factor, must be shown. But if one must be shown, the other must also be.

Examples:

$$1^o + 23^x \cdot 5 + 1^- \cdot 1 = 1^o + 23^x \cdot 5 + 1^-$$

$$1^o + 6^- \cdot 0 + (1^* + 1^-) = 1^o + (1^* + 1^-) \, .$$

16) If there are $k_1$ branches $n_{\widetilde{1}}$ (number of forms $Z_1$ per one branch), $k_2$ branches $n_{\widetilde{2}}$ (number of forms $Z_2$ per one branch), ... , $k_m$ branches $n_{\widetilde{m}}$ (number of forms $Z_m$ per one branch), then the whole quantity Z of sum forms on the row is

$$Z = \prod_{i=1}^{m} \binom{Z_i + k_i - 1}{k_i} \qquad (8)$$

Example:

The row $1^o + 4^{1.-4.} \cdot 2 + 3^{1.-2.} \cdot 3 + 2^- \cdot 10 + 1^-$ has the following number Z:

$$Z = \binom{4+2-1}{2}\binom{2+3-1}{3}\binom{1+10-1}{10}\binom{1+1-1}{1}$$

$$= \binom{5}{2}\binom{4}{3}\binom{10}{10}\binom{1}{1}$$

$$= \frac{5 \cdot 4 \cdot (3)!}{2!\,(5-2)!} \cdot \frac{4 \cdot (3)!}{3!\,(4-3)!} \cdot \frac{10!}{10!\,(0)!} \cdot \frac{1!}{1!\,(0)!}$$

$$= \frac{20 \cdot (3)!}{2 \cdot (3)!} \cdot \frac{4 \cdot (3)!}{1!\,(3)!} \cdot \frac{10!}{10!} \cdot \frac{1}{1} = \frac{20}{2} \cdot \frac{4}{1} \cdot 1 \cdot 1 = 10 \cdot 4$$

$$= 40.$$

## 1.2.2.3 Known graphical equalities

1) $1^{0-} + 1^{-} \cdot n \cong 1^{0-} + n^{x}$ ; $n \geq 3$

2) $1^{0-} + 1^{-} \cdot 1 \cong 1^{0-} + 1^{-}$

3) $1^{0-} + 1^{-} \cdot 2 \cong 1^{0-} + 2^{-}$

4) $1^{0-} + n^{-} \cong 1^{0-} + (a)^{-} + (n - a)^{-}$
$; n \geq 1, 0 \leq a < n$

5) $1^{0-} + (1^{*} + \Sigma_{1}) + \Sigma_{2} \cong 1^{0-} + (1^{*} + \Sigma_{2}) + \Sigma_{1}$

; where $\Sigma_{1}$, $\Sigma_{2}$ are arbitrary collections consisting of branches and external vertices. Examples could be:

$\Sigma_{1} = 3^{x} + 2^{-} \cdot 2 + 1^{-}$

$\Sigma_{2} = 15^{-} + (1^{*} + 5^{x} + (1^{*} + 1^{-} \cdot 3) + 1^{-}) + 2^{-}$

6) $1^{o-} + (1^* + n^\sim) + m^\sim \cong 1^{o-} + (1^* + m^\sim) + n^\sim$

7) $1^{o-} + n^{mL} \cong 1^{o-} + 2^- \cdot m + 1^- \cdot (n - 2m)$ ; $n \geq 2m + 1, n \geq 3, m \geq 0$

8) $1^{o-} + (2n + 1)^{nL} \cong 1^{o-} + 2^- \cdot n + 1^-$ ; $n \geq 1$

9) $1^{o-} + n^{1L} \cong 1^{o-} + (n - 1)^x + 1^-$ ; $n \geq 4$

10) $1^{o-} + n^Y \cong 1^{o-} + (1^* + (n - 2)^- + 1^-)$ ; $n \geq 4$

11) $1^{o-} + n^Y + m^- \cong 1^{o-} + (n + m)^Y$ ; $n \geq 4, m \geq 1$

12) $1^{o-} + 3^x + n^- \cong 1^{o-} + (n + 3)^Y$ ; $n \geq 1$

13) $1^{o-} + n^- + 1^- \cdot 2 \cong 1^{o-} + (n + 2)^Y$ ; $n \geq 2$

14) $1^{0-} + 2^- \cdot n + 1^- \cong 1^{0-} + (1^* + (1^* + 2^- \cdot (n-1) + 1^-)); n \geq 1$

15) $1^{0-} + (1^* + 0^{()}) \cdot n \cong 1^{0-} + 1^- \cdot n$
$; n \geq 1$

16) $1^{0-} + (1^* + 0^{()}) \cong 1^{0-} + 1^-$

17) $1^{0-} + 0^- \cong 1^{0-} + 0^x \cong 1^{0-} + 0^{()} \cong 1^{0-} + 0^{mL} \cong 1^{0-} + 0^Y \cong 1^{0-}$

18) $1^{0-} + 0^{()} \cdot 0 \cong 1^{0-}$

19) $1^{0-} + n^{\sim} \cdot 0 \cong 1^{0-}$

20) $1^{0-} + n^{\sim} \cdot 1 \cong 1^{0-} + n^{\sim}$.

The case 6) is a special case of the case 5). In the case 6) the branches $n^{\sim}, m^{\sim}$ can have any of their forms, and the cases $n^{\sim} = m^{\sim}$ and $n = m$ are possible (in the case $n = m$ the branches can still have different forms). In the cases 19) and 20) the branch $n^{\sim}$ can also have any of its forms.

The branches $n^{mL}$ and $n^Y$ are quite experimental concepts, as their first use is in this book. In this tget section they will not further be mentioned and were mentioned in the first place to give further options and research curiosities in vertex edge algebra. Searching for further graphical equalities in vertex edge algebra should also be a research option in its own regard, as also should any other concept in vertex edge algebra and the algorithms and also the condition matrix.

### 1.2.3Tree generation algorithm and its variations

The tree generation algorithm is an algorithm, which uses vertex edge algebra as its language. Another option would be to use partition language (sums of numbers). The algorithm can be used without computers to generate and enumerate trees, that may give an opportunity compared to other machine-based algorithms. By analyzing the outputs of the tree generation algorithm and its variations one can get deep insights into how an efficient and working tree generation and enumeration system works.

The tree generation algorithm's process for rooted trees is the following:

## 1.2.3.1Tree generation algorithm

1) Let $v = n$.

2) Set the root as a base, with removing number one (1) from the number $n$. The root is the first number in the partition and does not take part in the partitions.

3) Solve the partition equation

$$\sum_{i=1}^{m}(x_i) = n - 1 \, , \; m \leq n - 1 \, , \; x_i \in [1, n-1]$$
(9)

, with inserting each partition into its own column with the exception that the sum forms that have the same size largest branches go to the same columns. The order in the columns is from up to down, and the sum forms with the largest column specific branch $n^\sim$ are in the column $C_n$ , when the column's $C_1$ largest sum form object is an external vertex.

4) Repeat the part 3) with setting the internal vertices, the first number ones (1) inside brackets (branches), as bases, and then solving the partition equation. This happens with removing number one (1) from the size of the branch $p\tilde{}$ (size is greater than two, nothing happens to size two branch), and then solving the partition equation

$$\sum_{i=1}^{m}(x_i) = p - 1 \, , \ m \leq p - 1 \, , \ x_i \in [1, p - 1]$$

Remember that the number ones set as bases do not take part in the partitions. The whole sum forms that contain these partitions, are put into the same columns where the original sum forms were. If there are branches with different sizes, then the forms of the largest ones are gone through first. If there are branches with same sizes, then it does not matter which forms of a particular branch are gone through first, if the following example is considered.

Let there be a sum form, which has branches 3~ (two forms), 3~ (two forms) and 4~ (four forms). Then the combinations are the following:

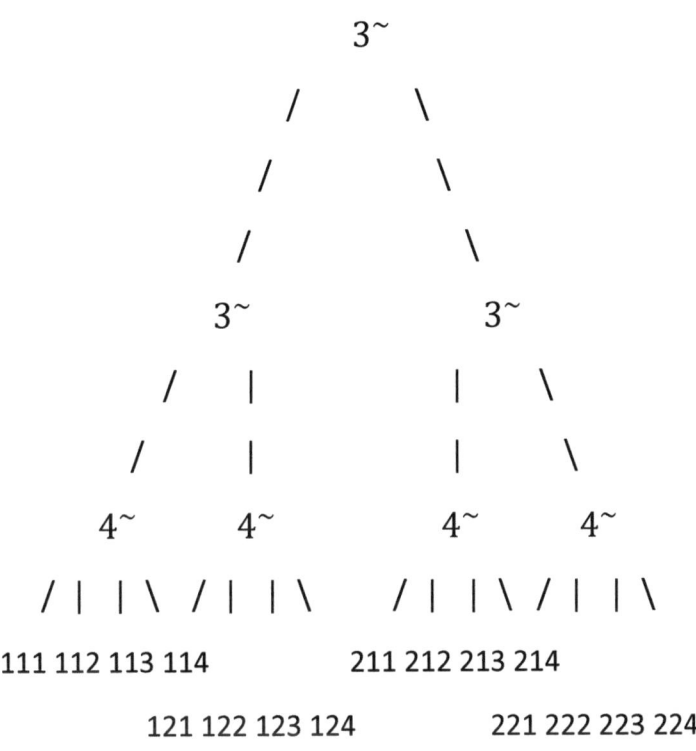

111 112 113 114          211 212 213 214

          121 122 123 124          221 222 223 224

Thus, if the sum form has a form $1^0 + 4^x + 3^x \cdot 2$ (code 111), then the following sum forms for the tree generation algorithm case would be the following, when the codes 21X are discarded due to similarity:

Code 111: $1^0 + 4^x + 3^x \cdot 2$

Code 112: $1^0 + 4^{2.} + 3^x \cdot 2$

Code 113: $1^0 + 4^{3.} + 3^x \cdot 2$

Code 114: $1^0 + 4^{4.} + 3^x \cdot 2$

Code 121: $1^0 + 4^{1.} + 3^- + 3^x$

Code 122: $1^0 + 4^{2.} + 3^- + 3^x$

Code 123: $1^0 + 4^{3.} + 3^- + 3^x$

Code 124: $1^0 + 4^{4.} + 3^- + 3^x$

Code 221: $1^0 + 4^{1.} + 3^- \cdot 2$

Code 222: $1^0 + 4^{2.} + 3^- \cdot 2$

Code 223: $1^0 + 4^{3.} + 3^- \cdot 2$

Code 224: $1^0 + 4^{4.} + 3^- \cdot 2$

$$4^{1.} = 4^x$$
$$4^{2.} = (1^* + 2^- + 1^-)$$
$$4^{3.} = (1^* + 3^x)$$
$$4^{4.} = 4^-$$

5) Repeat the part 4) until every partition has been gone through lexicographically and no further partitions are possible. Link every partition with inserting an equality sign between two consecutive partitions. Also, link the first partition with the number n (forest).

6) Change all the symbols in the following way:

6.1) The first number n, which has all the partitions, is the closed forest with a tree size n. The first equality sign is the symbol ≡.

6.2) The first number one (1) outside brackets is the root, the last number ones (inside and outside brackets) are external vertices and the first number one inside brackets (every starting bracket pair) is an internal vertex.

6.3) Every other equality sign is a vertex sum equivalence sign. The numbers on the right side of the multiplication sign $\cdot$ are quantity factors and the numbers on the left side of the multiplication sign are branches / external vertices.

6.4) The number two is a branch $2^-$ and the number k (k $\geq$ 3) is a branch $k^x$. The objects
$$(1 + 1), (1 + (1 + 1)) = (1 + 2), \left(1 + (1 + (1 + 1))\right) = (1 + (1 + 2)), \ldots,$$
can be compactified with the notions $2^-, 3^-, 4^-, \ldots$

.

# 1.2.3.2Examples:

i.      Let there be a rooted tree forest with tree size $v = 5$. The tree generation algorithm works as in the procedure. The partitions are already linked with equality signs in this and the following examples.

1) $v = 5$

2) $5 = 1 + 4 = 1 + 4 = 1 + 4 = 1 + 4$

$$= 1 + 4$$

3) $5 = 1 + 1 \cdot 4 = 1 + 2 + 1 \cdot 2$

$$= 1 + 2 \cdot 2$$

$$= 1 + 3 + 1$$

$$= 1 + 4 \, .$$

4) $5 = 1 + 1 \cdot 4 = 1 + 2 + 1 \cdot 2$
$$= 1 + 2 \cdot 2$$

$$= 1 + (1 + 1 \cdot 2) + 1$$

$$= 1 + (1 + 2) + 1$$

$$= 1 + (1 + 1 \cdot 3)$$
$$= 1 + (1 + 2 + 1)$$
$$= 1 + (1 + 3) \,.$$

5) $5 = 1 + 1 \cdot 4 = 1 + 2 + 1 \cdot 2$
$$= 1 + 2 \cdot 2$$

$$= 1 + (1 + 1 \cdot 2) + 1$$
$$= 1 + (1 + 2) + 1$$

$$= 1 + (1 + 1 \cdot 3)$$
$$= 1 + (1 + 2 + 1)$$
$$= 1 + \left(1 + (1 + 1 \cdot 2)\right)$$
$$= 1 + \left(1 + (1 + 2)\right) \,.$$

6) $\{5\}_{rie+}^{\Sigma o} \equiv 1^o + 1^- \cdot 4$

$$\doteq 1^o + 2^- + 1^- \cdot 2$$
$$\doteq 1^o + 2^- \cdot 2$$

$$\doteq 1^o + 3^x + 1^-$$
$$\doteq 1^o + 3^- + 1^-$$

$$\doteq 1^o + 4^x$$
$$\doteq 1^o + (1^* + 2^- + 1^-)$$
$$\doteq 1^o + (1^* + 3^x)$$
$$\doteq 1^o + 4^- \ .$$

ii.    Another rooted trees example, with tree size v = 6. The tree generation algorithm works as in the process.

1) $v = 6$

2) $6 = 1 + 5 = 1 + 5$

$$= 1 + 5$$

$$= 1 + 5$$
$$= 1 + 5$$

$$= 1 + 5$$

$$= 1 + 5$$

3) $6 = 1 + 1 \cdot 5 = 1 + 2 + 1 \cdot 3$
$$= 1 + 2 \cdot 2 + 1$$
$$= 1 + 3 + 1 \cdot 2$$

$$= 1 + 3 + 2$$

$$= 1 + 4 + 1$$

$$= 1 + 5 \,.$$

4) and 5) $6 = 1 + 1 \cdot 5 = 1 + 2 + 1 \cdot 3$
$$= 1 + 2 \cdot 2 + 1$$

$$= 1 + (1 + 1 \cdot 2) + 1 \cdot 2$$
$$= 1 + (1 + 2) + 1 \cdot 2$$
$$= 1 + (1 + 1 \cdot 2) + 2$$
$$= 1 + (1 + 2) + 2$$

$$= 1 + (1 + 1 \cdot 3) + 1$$
$$= 1 + (1 + 2 + 1) + 1$$
$$= 1 + \big(1 + (1 + 1 \cdot 2)\big) + 1$$
$$= 1 + \big(1 + (1 + 2)\big) + 1$$

$$= 1 + (1 + 1 \cdot 4)$$

$$= 1 + (1 + 2 + 1 \cdot 2)$$

$$= 1 + (1 + 2 \cdot 2)$$

$$= 1 + (1 + (1 + 1 \cdot 2) + 1)$$

$$= 1 + (1 + (1 + 2) + 1)$$

$$= 1 + \left(1 + (1 + 1 \cdot 3)\right)$$

$$= 1 + \left(1 + (1 + 2 + 1)\right)$$

$$= 1 + \left(1 + \left(1 + (1 + 1 \cdot 2)\right)\right)$$

$$= 1 + \left(1 + \left(1 + (1 + 2)\right)\right).$$

6) $\{6\}_{rie+}^{\Sigma o} \equiv 1^o + 1^- \cdot 5 \doteq 1^o + 2^- + 1^- \cdot 3$

$$\doteq 1^o + 2^- \cdot 2 + 1^-$$

$$\doteq 1^o + 3^x + 1^- \cdot 2$$

$$\doteq 1^o + 3^- + 1^- \cdot 2$$

$$\doteq 1^o + 3^x + 2^-$$

$$\doteq 1^o + 3^- + 2^-$$

$$\doteq 1^o + 4^x + 1^-$$

$$\doteq 1^o + (1^* + 2^- + 1^-) + 1^-$$

$$\doteq 1^o + (1^* + 3^x) + 1^-$$

$$\doteq 1^o + 4^- + 1^-$$

$$\doteq 1^o + 5^x$$

$$\doteq 1^o + (1^* + 2^- + 1^- \cdot 2)$$

$$\doteq 1^o + (1^* + 2^- \cdot 2)$$

$$\doteq 1^o + (1^* + 3^x + 1^-)$$

$$\doteq 1^o + (1^* + 3^- + 1^-)$$

$$\doteq 1^o + (1^* + 4^x)$$

$$\doteq 1^o + \left(1^* + (1^* + 2^- + 1^-)\right)$$

$$\doteq 1^o + \left(1^* + (1^* + 3^x)\right)$$

$$\doteq 1^o + 5^- .$$

iii.    Series-reduced rooted trees example. The tree generation algorithm works now differently:

a) The type $n^-$, $n \geq 2$ , branches are forbidden.

b) Outside brackets the number of objects (branch, external vertex) connected to the root must not be equal to two.

c) In a branch there must be at least two objects (branch, external vertex) connected to the internal vertex inside brackets. Thus, there must be three or more objects connected to the internal vertex in total, when the base connection point for the branch is taken into account. This rule applies for every internal vertex and bracket pair.

1) $v = 6$

2) $6 = 1 + 5 = 1 + 5 = 1 + 5$

3) , 4) and 5)

$$6 = 1 + 1 \cdot 5 = 1 + 3 + 1 \cdot 2$$

$$= 1 + (1 + 1 \cdot 4)$$
$$= 1 + (1 + 3 + 1) .$$

6) $\{6\}_{rie+}^{\Sigma o} \equiv 1^o + 1^- \cdot 5 \doteq 1^o + 3^x + 1^- \cdot 2$

$\doteq 1^o + 5^x$

$\doteq 1^o + (1^* + 3^x + 1^-) .$

iv.   Free trees example. The same rules as for
      the rooted trees, but with the following
      exceptions:

      a) Let $v = i + j = n > 5$; $i \leq j$; i and j
         are positive nonzero integers. The cases
         $v = 1 - 5$ will be shown after the parts
         a), b) and c).

      b) The columns j are discarded.

      c) In the case $v = n = 2p = i + j$ , in the
         column $i = j = p$:

         Let $Z_a$ be the number of forms for a
         branch $a\tilde{}$. Let k be a positive nonzero
         integer. Thus

         $$\vec{k!} = \sum_{m=0}^{k-1}(k - m) \quad (10)$$

         and

         $$\overleftarrow{k!} = \sum_{m=1}^{k}(m) \quad (11)$$

In the column $i = j = p$ the sum forms are accepted term by term due to the sum $(Z_p)\vec{!}$ and discarded term by term due to the sum $(Z_p - 1)\overleftarrow{!}$, in the following way:

$Z_p$ *accepted*, 1 *discarded*,

$Z_p - 1$ *accepted*,

2 *discarded*,

$Z_p - 2$ *accepted*,   3 *discarded*,

$Z_p - 3$ *accepted*, ... ,
$Z_p - 3$ *discarded*,
3 *accepted*,

$Z_p - 2$ *discarded*, 2 *accepted*,

$Z_p - 1$ *discarded*, 1 *accepted*.

$$\{1\}_{rie-}^{\Sigma o} \equiv 1^o$$

$$\{2\}_{rie-}^{\Sigma o} \equiv 1^o + 1^-$$

$$\{3\}_{rie-}^{\Sigma o} \equiv 1^o + 1^- \cdot 2$$

$$\{4\}_{rie-}^{\Sigma o} \equiv 1^o + 1^- \cdot 3 \doteq 1^o + 2^- + 1^-$$

$$\{5\}_{rie-}^{\Sigma o} \equiv 1^o + 1^- \cdot 4 \doteq 1^o + 2^- + 1^- \cdot 2$$
$$\doteq 1^o + 2^- \cdot 2$$

Notice that

$$\vec{k!} + (k-1)\overleftarrow{!} = k^2 \quad (12)$$

and

$$\vec{k!} = \frac{k(k+1)}{2} \quad (13)$$

1) $v = 8$

2) $8 = 1 + 7 = 1 + 7 = 1 + 7 = 1 + 7$

3) , 4) and 5)

$8 = 1 + 1 \cdot 7 = 1 + 2 + 1 \cdot 5$
$$= 1 + 2 \cdot 2 + 1 \cdot 3$$
$$= 1 + 2 \cdot 3 + 1$$
$$= 1 + (1 + 1 \cdot 2) + 1 \cdot 4$$
$$= 1 + (1 + 2) + 1 \cdot 4$$
$$= 1 + (1 + 1 \cdot 2) + 2 + 1 \cdot 2$$
$$= 1 + (1 + 2) + 2 + 1 \cdot 2$$
$$= 1 + (1 + 1 \cdot 2) + 2 \cdot 2$$
$$= 1 + (1 + 2) + 2 \cdot 2$$
$$= 1 + (1 + 1 \cdot 2) \cdot 2 + 1$$
$$= 1 + (1 + 2) + (1 + 1 \cdot 2) + 1$$
$$= 1 + (1 + 2) \cdot 2 + 1$$

$$= 1 + (1 + 1 \cdot 3) + 1 \cdot 3$$

$$= 1 + (1 + 2 + 1) + 1 \cdot 3$$

$$= 1 + \big(1 + (1 + 1 \cdot 2)\big) + 1 \cdot 3$$

$$= 1 + \big(1 + (1 + 2)\big) + 1 \cdot 3$$

$$= 1 + (1 + 2 + 1) + 2 + 1$$

$$= 1 + \big(1 + (1 + 1 \cdot 2)\big) + 2 + 1$$

$$= 1 + \big(1 + (1 + 2)\big) + 2 + 1$$

$$= 1 + \big(1 + (1 + 1 \cdot 2)\big) + (1 + 1 \cdot 2)$$

$$= 1 + \big(1 + (1 + 2)\big) + (1 + 1 \cdot 2)$$

$$= 1 + \big(1 + (1 + 2)\big) + (1 + 2) \, .$$

6) $\{8\}_{rie-}^{\Sigma o} \equiv 1^o + 1^- \cdot 7 \doteq 1^o + 2^- + 1^- \cdot 5$

$$\doteq 1^o + 2^- \cdot 2 + 1^- \cdot 3$$

$$\doteq 1^o + 2^- \cdot 3 + 1^-$$

$$\doteq 1^o + 3^x + 1^- \cdot 4$$

$$\doteq 1^o + 3^- + 1^- \cdot 4$$

$$\doteq 1^o + 3^x + 2^- + 1^- \cdot 2$$

$$\doteq 1^o + 3^- + 2^- + 1^- \cdot 2$$

$$\doteq 1^o + 3^x + 2^- \cdot 2$$

$$\doteq 1^o + 3^- + 2^- \cdot 2$$

$$\doteq 1^o + 3^x \cdot 2 + 1^-$$

$$\doteq 1^o + 3^- + 3^x + 1^-$$

$$\doteq 1^o + 3^- \cdot 2 + 1^-$$


$$\doteq 1^o + 4^x + 1^- \cdot 3$$

$$\doteq 1^o + (1^* + 2^- + 1^-) + 1^- \cdot 3$$

$$\doteq 1^o + (1^* + 3^x) + 1^- \cdot 3$$

$$\doteq 1^o + 4^- + 1^- \cdot 3$$

$$\doteq 1^o + (1^* + 2^- + 1^-) + 2^- + 1^-$$

$$\doteq 1^o + (1^* + 3^x) + 2^- + 1^-$$

$$\doteq 1^o + 4^- + 2^- + 1^-$$

$$\doteq 1^o + (1^* + 3^x) + 3^x$$

$$\doteq 1^o + 4^- + 3^x$$

$$\doteq 1^o + 4^- + 3^-.$$

v.  Series-reduced free trees example. The tree generation algorithm works the same as for the rooted trees case, but with additional conditions from both the series-reduced rooted trees case and free trees case.

1) v = 10

2) $10 = 1 + 9 = 1 + 9 = 1 + 9 = 1 + 9$

3) , 4) and 5)

$$10 = 1 + 1 \cdot 9 = 1 + 3 + 1 \cdot 6$$
$$= 1 + 3 \cdot 2 + 1 \cdot 3$$
$$= 1 + 3 \cdot 3$$

$$= 1 + 4 + 1 \cdot 5$$
$$= 1 + 4 + 3 + 1 \cdot 2$$
$$= 1 + 4 \cdot 2 + 1$$

$$= 1 + (1 + 1 \cdot 4) + 1 \cdot 4$$
$$= 1 + (1 + 3 + 1) + 1 \cdot 4$$
$$= 1 + (1 + 3 + 1) + 3 + 1$$

$$6)\{10\}_{rie-}^{\Sigma o} \equiv 1^o + 1^- \cdot 9$$

$$\doteq 1^o + 3^x + 1^- \cdot 6$$

$$\doteq 1^o + 3^x \cdot 2 + 1^- \cdot 3$$

$$\doteq 1^o + 3^x \cdot 3$$

$$\doteq 1^o + 4^x + 1^- \cdot 5$$

$$\doteq 1^o + 4^x + 3^x + 1^- \cdot 2$$

$$\doteq 1^o + 4^x \cdot 2 + 1^-$$

$$\doteq 1^o + 5^x + 1^- \cdot 4$$

$$\doteq 1^o + (1^* + 3^x + 1^-) + 1^- \cdot 4$$

$$\doteq 1^o + (1^* + 3^x + 1^-) + 3^x + 1^-$$

### 1.2.3.3 Information about the tree generation algorithm

The tree generation algorithm is based on partitions: in how many ways can the number $v - 1$ be expressed as sums of positive nonzero integers, when v is the size of a tree. Sum forms, which can be expressed by the plain language of partitions, correspond to ordinary partitions of the number $v - 1$, with the except for the root; The sum forms with brackets are an additional feature of trees, and are not specifically ordinary partitions of the number $v - 1$. However, the brackets do not in an arithmetical way change these partitions, so the notion that trees are partitions is still quite right.

What comes to depth levels, the root corresponds to first and internal vertices to second, third and so on. The depth level starters ensure that every combination is considered, without the possibility of accidentally changing the places of branches during the generation process. An example of depth levels (the number in the exponent of the root / internal vertex defines the depth level):

$$(17)^o_{rie+} = 1^{01} + (1^{*2} + 3^x \cdot 2 + (1^{*3} + 1^- \cdot 4) + 2^-) + (1^{*2} + 1^-)$$

Notice that the depth levels are the same for two internal vertices when these are inside the same number of bracket pairs.

The generated forest is divided into columns. The column $C_n$ defines the size of the largest branch $n\tilde{\ }$ in the sum form. This way for example the column $C_2$ never exists for the series-reduced rooted and series-reduced free trees cases. The dividing of the columns according to the largest branch in the sum form eases the construction of forests. In these columns the partitions are put into lexicographical order (from up to down), thus creating a working generation process.

## 1.2.3.4 Branch form generator and tree enumeration algorithm

Branch form generator is a recursive algorithm, which generates all the forms of the chosen branch $n^{\sim}$ , thus enumerating the number $Z_n$, to be used by the tree enumeration algorithm. The tree enumeration algorithm, on the other hand, works the same as tree generation algorithm, but with some exceptions.

## 1.2.3.5 Branch form generator

1) Let there be a branch $n^{\sim}$.

2) Think of the forms of the branch being in the same column. The generation order is from up to down and a new form starts a new row. Act in the way defined by the tree generation algorithm and use lexicographic order.

3) Use the notation $m^{a.-b.}$ as much as possible.

4) Use the notation $2^-$(a,b)(rooted and free trees) and the notation $3^x$(a,b)(series-reduced rooted trees and series-reduced free trees) as much as possible.

## 1.2.3.6Tree enumeration algorithm

1) The tree enumeration algorithm works the same as the tree generation algorithm with the exception that the notations $m^{a.-b.}$, $2^-$(a,b) and $3^x$(a,b) are used as much as possible.

2) Combination coefficients (Z) come in brackets after every enumeration row.

## 1.2.3.7Examples

i.     Rooted trees, v = 1 − 10. For this tree type $Z_v = r_v$ , so one algorithm is enough.

$1^{1.} = 1^-$

$2^{1.} = 2^-$

$3^{1.-2.} = (1^* + 2^-(0,1) + 1^-(2,0))$

$4^{1.-2.} = (1^* + 2^-(0,1) + 1^-(3,1))$

$4^{3.-4.} = (1^* + 3^{1.-2.})$

$5^{1.-3.} = (1^* + 2^-(0,2) + 1^-(4,0))$

$5^{4.-5.} = (1^* + 3^{1.-2.} + 1^-)$

$5^{6.-9.} = (1^* + 4^{1.-4.})$

$6^{1.-3.} = (1^* + 2^-(0,2) + 1^-(5,1))$

$6^{4.-7.} = (1^* + 3^{1.-2.} + 2^-(0,1) + 1^-(2,0))$

$6^{8.-11.} = (1^* + 4^{1.-4.} + 1^-)$

$6^{12.-20.} = (1^* + 5^{1.-9.})$

$$7^{1.-4.} = (1^* + 2^-(0,3) + 1^-(6,0))$$

$$7^{5.-8.} = (1^* + 3^{1.-2.} + 2^-(0,1) + 1^-(3,1))$$

$$7^{9.-11.} = (1^* + 3^{1.-2.} \cdot 2)$$

$$7^{12.-19.} = (1^* + 4^{1.-4.} + 2^-(0,1) + 1^-(2,0))$$

$$7^{20.-28.} = (1^* + 5^{1.-9.} + 1^-)$$
$$7^{29.-48.} = (1^* + 6^{1.-20.})$$
$$8^{1.-4.} = (1^* + 2^-(0,3) + 1^-(7,1))$$
$$8^{5.-10.} = (1^* + 3^{1.-2.} + 2^-(0,2) + 1^-(4,0))$$
$$8^{11.-13.} = (1^* + 3^{1.-2.} \cdot 2 + 1^-)$$
$$8^{14.-21.} = (1^* + 4^{1.-4.} + 2^-(0,1) + 1^-(3,1))$$
$$8^{22.-29.} = (1^* + 4^{1.-4.} + 3^{1.-2.})$$
$$8^{30.-47.} = (1^* + 5^{1.-9.} + 2^-(0,1) + 1^-(2,0))$$
$$8^{48.-67.} = (1^* + 6^{1.-20.} + 1^-)$$
$$8^{68.-115.} = (1^* + 7^{1.-48.})$$
$$9^{1.-5.} = (1^* + 2^-(0,4) + 1^-(8,0))$$
$$9^{6.-11.} = (1^* + 3^{1.-2.} + 2^-(0,2) + 1^-(5,1))$$
$$9^{12.-17.} = (1^* + 3^{1.-2.} \cdot 2 + 2^-(0,1) +$$
$$1^-(2,0))$$
$$9^{18.-29.} = (1^* + 4^{1.-4.} + 2^-(0,2) + 1^-(4,0))$$
$$9^{30.-37.} = (1^* + 4^{1.-4.} + 3^{1.-2.} + 1^-)$$
$$9^{38.-47.} = (1^* + 4^{1.-4.} \cdot 2)$$
$$9^{48.-65.} = (1^* + 5^{1.-9.} + 2^-(0,1) + 1^-(3,1))$$
$$9^{66.-83.} = (1^* + 5^{1.-9.} + 3^{1.-2.})$$

$9^{84.-123.} = (1^* + 6^{1.-20.} + 2^-(0,1) +$
$1^-(2,0))$

$9^{124.-171.} = (1^* + 7^{1.-48.} + 1^-)$

$9^{172.-286.} = (1^* + 8^{1.-115.})$

$10^{1.-5.} = (1^* + 2^-(0,4) + 1^-(9,1))$

$10^{6.-13.} = (1^* + 3^{1.-2.} + 2^-(0,3) + 1^-(6,0))$

$10^{14.-19.} = (1^* + 3^{1.-2.} \cdot 2 + 2^-(0,1) +$
$1^-(3,1))$

$10^{20.-23.} = (1^* + 3^{1.-2.} \cdot 3)$

$10^{24.-35.} = (1^* + 4^{1.-4.} + 2^-(0,2) +$
$1^-(5,1))$

$10^{36.-51.} = (1^* + 4^{1.-4.} + 3^{1.-2.} + 2^-(0,1) +$
$1^-(2,0))$

$10^{52.-61.} = (1^* + 4^{1.-4.} \cdot 2 + 1^-)$

$10^{62.-88.} = (1^* + 5^{1.-9.} + 2^-(0,2) +$
$1^-(4,0))$

$10^{89.-106.} = (1^* + 5^{1.-9.} + 3^{1.-2.} + 1^-)$

$10^{107.-142.} = (1^* + 5^{1.-9.} + 4^{1.-4.})$

$10^{143.-182.} = (1^* + 6^{1.-20.} + 2^-(0,1) +$
$1^-(3,1))$

$10^{183.-222.} = (1^* + 6^{1.-20.} + 3^{1.-2.})$

$10^{223.-318.} = (1^* + 7^{1.-48.} + 2^-(0,1) +$
$1^-(2,0))$

$10^{319.-433.} = (1^* + 8^{1.-115.} + 1^-)$

$10^{434.-719.} = (1^* + 9^{1.-286.})$

Thus

| v: | 1 | 2 | 3 | 4 | 5 |
|---|---|---|---|---|---|
| $r_v$: | 1 | 1 | 2 | 4 | 9 |
| Number of algorithm rows: | 1 | 1 | 1 | 2 | 3 |

| v: | 6 | 7 | 8 | 9 | 10 |
|---|---|---|---|---|---|
| $r_v$: | 20 | 48 | 115 | 286 | 719 |
| Number of algorithm rows: | 4 | 6 | 8 | 11 | 15 |

ii. Series-reduced rooted trees, v = 1 –
10. Now also the tree enumeration
algorithm is needed.

$1^{1.} = 1^-$

$3^{1.} = 3^x$

$4^{1.} = 4^x$

$5^{1.-2.} = (1^* + 3^x(0,1) + 1^-(4,1))$

$6^{1.-2.} = (1^* + 3^x(0,1) + 1^-(5,2))$

$6^{3.} = (1^* + 4^x + 1^-)$

$7^{1.-3.} = (1^* + 3^x(0,2) + 1^-(6,0))$

$7^{4.} = (1^* + 4^x + 1^- \cdot 2)$

$7^{5.-6.} = (1^* + 5^{1.-2.} + 1^-)$

$8^{1.-3.} = (1^* + 3^x(0,2) + 1^-(7,1))$

$8^{4.-5.} = (1^* + 4^x + 3^x(0,1) + 1^-(3,0))$

$8^{6.-7.} = (1^* + 5^{1.-2.} + 1^- \cdot 2)$

$$8^{8.-10.} = (1^* + 6^{1.-3.} + 1^-)$$

$$9^{1.-3.} = (1^* + 3^x(0,2) + 1^-(8,2))$$

$$9^{4.-5.} = (1^* + 4^x + 3^x(0,1) + 1^-(4,1))$$

$$9^{6.} = (1^* + 4^x \cdot 2)$$

$$9^{7.-10.} = (1^* + 5^{1.-2.} + 3^x(0,1) + 1^-(3,0))$$

$$9^{11.-13.} = (1^* + 6^{1.-3.} + 1^- \cdot 2)$$

$$9^{14.-19.} = (1^* + 7^{1.-6.} + 1^-)$$

$$\{1\}_{rie+}^{\Sigma o} \equiv 1^o(Z = 1)$$

$$\{2\}_{rie+}^{\Sigma o} \equiv 1^o + 1^-(Z = 1)$$

$$\{3\}_{rie+}^{\Sigma o} \equiv [1^o + 1^- \cdot 2](Z = 0) \text{ (sum form}$$
discarded)

$$\{4\}_{rie+}^{\Sigma o} \equiv 1^o + 3^x(0,1) + 1^-(3,0)(Z = 2)$$

$$\{5\}_{rie+}^{\Sigma o} \equiv 1^o + 1^- \cdot 4(Z = 1) \doteq 1^o + 4^x(Z = 1)$$

$$\{6\}_{rie+}^{\Sigma o} \equiv 1^o + 3^x(0,1) + 1^-(5,2)(Z = 2)$$

$$\doteq 1^o + 5^{1.-2.}(Z = 2)$$

$$\{7\}_{rie+}^{\Sigma o} \equiv 1^o + 3^x(0,1) + 1^-(6,3)(Z = 2)$$

$$\doteq 1^o + 4^x + 1^- \cdot 2(Z = 1)$$

$$\doteq 1^o + 6^{1.-3.}(Z = 3)$$

$$\{8\}_{rie+}^{\Sigma o} \equiv 1^o + 3^x(0,2) + 1^-(7,1)(Z = 3)$$

$$\doteq 1^o + 4^x + 1^- \cdot 3(Z = 1)$$

$$\doteq 1^o + 5^{1.-2.} + 1^- \cdot 2(Z = 2)$$

$$\doteq 1^o + 7^{1.-6.}(Z = 6)$$

$$\{9\}^{\Sigma o}_{rie+} \equiv 1^o + 3^x(0,2) + 1^-(8,2)(Z = 3)$$

$$\doteq 1^o + 4^x + 3^x(0,1) + 1^-(4,1)(Z = 2)$$

$$\doteq 1^o + 5^{1.-2.} + 1^- \cdot 3(Z = 2)$$

$$\doteq 1^o + 6^{1.-3.} + 1^- \cdot 2(Z = 3)$$

$$\doteq 1^o + 8^{1.-10.}(Z = 10)$$

$$\{10\}^{\Sigma o}_{rie+} \equiv 1^o + 3^x(0,3) + 1^-(9,0)(Z = 4)$$

$$\doteq 1^o + 4^x + 3^x(0,1) + 1^-(5,2)(Z = 2)$$

$$\doteq 1^o + 4^x \cdot 2 + 1^-(Z = 1)$$

$$\doteq 1^o + 5^{1.-2.} + 3^x(0,1) + 1^-(4,1)(Z = 4)$$

$$\doteq 1^o + 6^{1.-3.} + 1^- \cdot 3(Z = 3)$$

$$\doteq 1^o + 7^{1.-6.} + 1^- \cdot 2(Z = 6)$$

$$\doteq 1^o + 9^{1.-19.}(Z = 19)$$

Thus

$$s_1 = 1$$

$$s_2 = 1$$

$$s_3 = 0$$

$$s_4 = 2$$

$$s_5 = 1 + 1 = 2$$

$$s_6 = 2 + 2 = 4$$

$$s_7 = 2 + 1 + 3 = 6$$

$$s_8 = 3 + 1 + 2 + 6 = 12$$

$$s_9 = 3 + 2 + 2 + 3 + 10 = 20$$

$$s_{10} = 4 + 2 + 1 + 4 + 3 + 6 + 19 = 39$$

iii.   Free trees, $v = 1 - 10$. The generated branch forms in the rooted trees case can now be used. The sum forms in the middle column are not compactified for the sake of clarity.

$$\{1\}_{rie-}^{\Sigma o} \equiv 1^o (Z = 1)$$

$$\{2\}_{rie-}^{\Sigma o} \equiv 1^o + 1^-(Z = 1)$$

$$\{3\}_{rie-}^{\Sigma o} \equiv 1^o + 1^- \cdot 2(Z = 1)$$

$$\{4\}_{rie-}^{\Sigma o} \equiv 1^o + 2^-(0,1) + 1^-(3,1)(Z = 2)$$

$$\{5\}_{rie-}^{\Sigma o} \equiv 1^o + 2^-(0,2) + 1^-(4,0)(Z = 3)$$

$$\{6\}^{\Sigma o}_{rie-} \equiv 1^o + 2^-(0,2) + 1^-(5,1)(Z = 3)$$

$$\doteq 1^o + 3^{1.-2.} + 1^- \cdot 2(Z = 2)$$

$$\doteq 1^o + 3^{2.} + 2^-(Z = 1)$$

$$\{7\}^{\Sigma o}_{rie-} \equiv 1^o + 2^-(0,3) + 1^-(6,0)(Z = 4)$$

$$\doteq 1^o + 3^{1.-2.} + 2^-(0,1) +$$
$$1^-(3,1)(Z = 4)$$

$$\doteq 1^o + 3^{1.-2.} \cdot 2(Z = 3)$$

$$\{8\}^{\Sigma o}_{rie-} \equiv 1^o + 2^-(0,3) + 1^-(7,1)(Z = 4)$$

$$\doteq 1^o + 3^{1.-2.} + 2^-(0,2) +$$
$$1^-(4,0)(Z = 6)$$

$$\doteq 1^o + 3^{1.-2.} \cdot 2 + 1^-(Z = 3)$$

$$\doteq 1^o + 4^{1.-4.} + 1^- \cdot 3(Z = 4)$$

$$\doteq 1^o + 4^{2.-4.} + 2^- + 1^-(Z = 3)$$

$$\doteq 1^o + 4^{3.-4.} + 3^{1.}(Z = 2)$$

$$\doteq 1^o + 4^{4.} + 3^{2.}(Z = 1)$$

$$\{9\}_{rie-}^{\Sigma o} \equiv 1^o + 2^-(0,4) + 1^-(8,0)(Z = 5)$$

$$\doteq 1^o + 3^{1.-2.} + 2^-(0,2) + 1^-(5,1)(Z = 6)$$

$$\doteq 1^o + 3^{1.-2.} \cdot 2 + 2^-(0,1) + 1^-(2,0)(Z = 6)$$

$$\doteq 1^o + 4^{1.-4.} + 2^-(0,2) + 1^-(4,0)(Z = 12)$$

$$\doteq 1^o + 4^{1.-4.} + 3^{1.-2.} + 1^-(Z = 8)$$

$$\doteq 1^o + 4^{1.-4.} \cdot 2(Z = 10)$$


$$\{10\}_{rie-}^{\Sigma o} \equiv 1^o + 2^-(0,4) + 1^-(9,1)(Z = 5)$$

$$\doteq 1^o + 3^{1.-2.} + 2^-(0,3) + 1^-(6,0)(Z = 8)$$

$$\doteq 1^o + 3^{1.-2.} \cdot 2 + 2^-(0,1) + 1^-(3,1)(Z = 6)$$

$$\doteq 1^o + 3^{1.-2.} \cdot 3(Z = 4)$$

$$\doteq 1^o + 4^{1.-4.} + 2^-(0,2) + 1^-(5,1)(Z =$$
12)

$$\doteq 1^o + 4^{1.-4.} + 3^{1.-2.} + 2^-(0,1) +$$
$$1^-(2,0)(Z = 16)$$

$$\doteq 1^o + 4^{1.-4.} \cdot 2 + 1^-(Z = 10)$$

$$\doteq 1^o + 5^{1.-9.} + 1^- \cdot 4(Z = 9)$$

$$\doteq 1^o + 5^{2.-9.} + 2^- + 1^- \cdot 2(Z = 8)$$

$$\doteq 1^o + 5^{3.-9.} + 2^- \cdot 2(Z = 7)$$

$$\doteq 1^o + 5^{4.-9.} + 3^{1.} + 1^-(Z = 6)$$

$$\doteq 1^o + 5^{5.-9.} + 3^{2.} + 1^-(Z = 5)$$

$$\doteq 1^o + 5^{6.-9.} + 4^{1.}(Z = 4)$$

$$\doteq 1^o + 5^{7.-9.} + 4^{2.}(Z = 3)$$

$$\doteq 1^o + 5^{8.-9.} + 4^{3.}(Z = 2)$$

$$\doteq 1^o + 5^{9.} + 4^{4.}(Z = 1)$$

Thus

$$t_1 = 1$$

$$t_2 = 1$$

$$t_3 = 1$$

$$t_4 = 2$$

$$t_5 = 3$$

$$t_6 = 3 + 2 + 1 = 6$$

$$t_7 = 4 + 4 + 3 = 11$$

$$t_8 = 4 + 6 + 3 + 4 + 3 + 2 + 1 = 23$$

$$t_9 = 5 + 6 + 6 + 12 + 8 + 10 = 47$$

$$t_{10} = 5 + 8 + 6 + 4 + 12 + 16 + 10 + 9 + \\ 8 + 7 + 6 + 5 + 4 + 3 + 2 + 1 = 106$$

iv. Series-reduced free trees, v = 1 −
10. The generated branch forms in
the series-reduced rooted trees
case can now be used. The sum
forms in the middle column are not
compactified for the sake of clarity.

$$\{1\}_{rie-}^{\Sigma o} \equiv 1^o(Z = 1)$$

$$\{2\}_{rie-}^{\Sigma o} \equiv 1^o + 1^-(Z = 1)$$

$$\{3\}_{rie-}^{\Sigma o} \equiv [1^o + 1^- \cdot 2](Z = 0)(\text{Sum}$$
form discarded)

$$\{4\}_{rie-}^{\Sigma o} \equiv 1^o + 1^- \cdot 3(Z = 1)$$

$$\{5\}_{rie-}^{\Sigma o} \equiv 1^o + 1^- \cdot 4(Z = 1)$$

$$\{6\}_{rie-}^{\Sigma o} \equiv 1^o + 3^x(0,1) + 1^-(5,2)(Z =$$
2)

$$\{7\}_{rie-}^{\Sigma o} \equiv 1^o + 3^x(0,1) + 1^-(6,3)(Z = 2)$$

$$\{8\}_{rie-}^{\Sigma o} \equiv 1^o + 3^x(0,2) + 1^-(7,1)(Z = 3)$$

$$\doteq 1^o + 4^x + 1^- \cdot 3(Z = 1)$$

$$\{9\}_{rie-}^{\Sigma o} \equiv 1^o + 3^x(0,2) + 1^-(8,2)(Z = 3)$$

$$\doteq 1^o + 4^x + 3^x(0,1) + 1^-(4,1)(Z = 2)$$

$$\{10\}_{rie-}^{\Sigma o} \equiv 1^o + 3^x(0,3) + 1^-(9,0)(Z = 4)$$

$$\doteq 1^o + 4^x + 3^x(0,1) + 1^-(5,2)(Z = 2)$$

$$\doteq 1^o + 4^x \cdot 2 + 1^-(Z = 1)$$

$$\doteq 1^o + 5^{1.-2.} + 1^- \cdot 4(Z = 2)$$

$$\doteq 1^o + 5^{2.} + 3^x + 1^-(Z = 1)$$

Thus

$$h_1 = 1$$

$$h_2 = 1$$

$$h_3 = 0$$

$$h_4 = 1$$

$$h_5 = 1$$

$$h_6 = 2$$

$$h_7 = 2$$

$$h_8 = 3 + 1 = 4$$

$$h_9 = 3 + 2 = 5$$

$$h_{10} = 4 + 2 + 1 + 2 + 1 = 10$$

Let us show now a more intensive enumeration example for series-reduced free trees.

$$1^{1.} = 1^-$$

$$3^{1.} = 3^x$$

$$4^{1.} = 4^x$$

$$5^{1.-2.} = (1^* + 3^x(0,1) + 1^-(4,1))$$

$$6^{1.-2.} = (1^* + 3^x(0,1) + 1^-(5,2))$$

$$6^{3.} = (1^* + 4^x + 1^-)$$

$$7^{1.-3.} = (1^* + 3^x(0,2) + 1^-(6,0))$$

$$7^{4.} = (1^* + 4^x + 1^- \cdot 2)$$

$$7^{5.-6.} = (1^* + 5^{1.-2.} + 1^-)$$

$$8^{1.-3.} = (1^* + 3^x(0,2) + 1^-(7,1))$$

$$8^{4.-5.} = (1^* + 4^x + 3^x(0,1) + 1^-(3,0))$$

$$8^{6.-7.} = (1^* + 5^{1.-2.} + 1^- \cdot 2)$$

$$8^{8.-10.} = (1^* + 6^{1.-3.} + 1^-)$$

$$9^{1.-3.} = (1^* + 3^x(0,2) + 1^-(8,2))$$

$$9^{4.-5.} = (1^* + 4^x + 3^x(0,1) + 1^-(4,1))$$

$$9^{6.} = (1^* + 4^x \cdot 2)$$

$$9^{7.-10.} = (1^* + 5^{1.-2.} + 3^x(0,1) + 1^-(3,0))$$

$$9^{11.-13.} = (1^* + 6^{1.-3.} + 1^- \cdot 2)$$

$$9^{14.-19.} = (1^* + 7^{1.-6.} + 1^-)$$

$$10^{1.-4.} = (1^* + 3^x(0,3) + 1^-(9,0))$$

$$10^{5.-6.} = (1^* + 4^x + 3^x(0,1) + 1^-(5,2))$$

$$10^{7.} = (1^* + 4^x \cdot 2 + 1^-)$$

$$10^{8.-11.} = (1^* + 5^{1.-2.} + 3^x(0,1) + 1^-(4,1))$$

$$10^{12.-13.} = (1^* + 5^{1.-2.} + 4^x)$$

$$10^{14.-19.} = (1^* + 6^{1.-3.} + 3^x(0,1) + 1^-(3,0))$$

$$10^{20.-25.} = (1^* + 7^{1.-6.} + 1^- \cdot 2)$$

$$10^{26.-35.} = (1^* + 8^{1.-10.} + 1^-)$$

v = 20.

As in the column $C_{20/2} = C_{10}$ there are $(Z_{10})\vec{!} = 35\vec{!} =$
$\frac{35\cdot(1+35)}{2} = \frac{35\cdot36}{2} = \frac{(30+5)(30+6)}{2} = \frac{900+180+150+30}{2} =$
$\frac{1260}{2} = 630$ sum forms, this column can be left out from
the next enumeration process, while the number 630 is
added to the result.

$\{20\}_{rie-}^{\Sigma o} \equiv 1^o + 3^x(0,6) + 1^-(19,1)(Z = 7)$

$\doteq 1^o + 4^x + 3^x(0,5) + 1^-(15,0)(Z = 6)$

$\doteq 1^o + 4^x \cdot 2 + 3^x(0,3) + 1^-(11,2)(Z = 4)$

$\doteq 1^o + 4^x \cdot 3 + 3^x(0,2) + 1^-(7,1)(Z = 3)$

$\doteq 1^o + 4^x \cdot 4 + 3^x(0,1) + 1^-(3,0)(Z = 2)$

$\doteq 1^o + 5^{1.-2.} + 3^x(0,4) + 1^-(14,2)(Z = 10)$

$\doteq 1^o + 5^{1.-2.} + 4^x + 3^x(0,3) + 1^-(10,1)(Z = 8)$

$\doteq 1^o + 5^{1.-2.} + 4^x \cdot 2 + 3^x(0,2) + 1^-(6,0)(Z = 6)$

$\doteq 1^o + 5^{1.-2.} + 4^x \cdot 3 + 1^- \cdot 2(Z = 2)$

$$\doteq 1^o + 5^{1.-2.} \cdot 2 + 3^x(0,3) + 1^-(9,0)(Z = 12)$$

$$\doteq 1^o + 5^{1.-2.} \cdot 2 + 4^x + 3^x(0,1) + 1^-(5,2)(Z = 6)$$

$$\doteq 1^o + 5^{1.-2.} \cdot 2 + 4^x \cdot 2 + 1^-(Z = 3)$$

$$\doteq 1^o + 5^{1.-2.} \cdot 3 + 3^x(0,1) + 1^-(4,1)(Z = 8)$$

$$\doteq 1^o + 5^{1.-2.} \cdot 3 + 4^x(Z = 4)$$

$$\doteq 1^o + 6^{1.-3.} + 3^x(0,4) + 1^-(13,1)(Z = 15)$$

$$\doteq 1^o + 6^{1.-3.} + 4^x + 3^x(0,3) + 1^-(9,0)(Z = 12)$$

$$\doteq 1^o + 6^{1.-3.} + 4^x \cdot 2 + 3^x(0,1) + 1^-(5,2)(Z = 6)$$

$$\doteq 1^o + 6^{1.-3.} + 4^x \cdot 3 + 1^-(Z = 3)$$

$$\doteq 1^o + 6^{1.-3.} + 5^{1.-2.} + 3^x(0,2) + 1^-(8,2)(Z = 18)$$

$$\doteq 1^o + 6^{1.-3.} + 5^{1.-2.} + 4^x + 3^x(0,1) + 1^-(4,1)(Z = 12)$$

$$\doteq 1^o + 6^{1.-3.} + 5^{1.-2.} + 4^x \cdot 2(Z = 6)$$

$$\doteq 1^o + 6^{1.-3.} + 5^{1.-2.} \cdot 2 + 3^x(0,1) + 1^-(3,0)(Z = 18)$$

$$\doteq 1^o + 6^{1.-3.} \cdot 2 + 3^x(0,2) + 1^-(7,1)(Z = 18)$$

$$\doteq 1^o + 6^{1.-3.} \cdot 2 + 4^x + 3^x(0,1) + 1^-(3,0)(Z = 12)$$

$$\doteq 1^o + 6^{1.-3.} \cdot 2 + 5^{1.-2.} + 1^- \cdot 2(Z = 12)$$

$$\doteq 1^o + 6^{1.-3.} \cdot 3 + 1^-(Z = 10)$$

$$\doteq 1^o + 7^{1.-6.} + 3^x(0,4) + 1^-(12,0)(Z = 30)$$

$$\doteq 1^o + 7^{1.-6.} + 4^x + 3^x(0,2) + 1^-(8,2)(Z = 18)$$

$$\doteq 1^o + 7^{1.-6.} + 4^x \cdot 2 + 3^x(0,1) + 1^-(4,1)(Z = 12)$$

$$\doteq 1^o + 7^{1.-6.} + 4^x \cdot 3(Z = 6)$$

$$\doteq 1^o + 7^{1.-6.} + 5^{1.-2.} + 3^x(0,2) + 1^-(7,1)(Z = 36)$$

$$\doteq 1^o + 7^{1.-6.} + 5^{1.-2.} + 4^x + 3^x(0,1) + 1^-(3,0)(Z = 24)$$

$$\doteq 1^o + 7^{1.-6.} + 5^{1.-2.} \cdot 2 + 1^- \cdot 2(Z = 18)$$

$$\doteq 1^o + 7^{1.-6.} + 6^{1.-3.} + 3^x(0,2) + 1^-(6,0)(Z = 54)$$

$$\doteq 1^o + 7^{1.-6.} + 6^{1.-3.} + 4^x + 1^- \cdot 2(Z = 18)$$

$$\doteq 1^o + 7^{1.-6.} + 6^{1.-3.} + 5^{1.-2.} + 1^-(Z = 36)$$

$$\doteq 1^o + 7^{1.-6.} + 6^{1.-3.} \cdot 2(Z = 36)$$

$$\doteq 1^o + 7^{1.-6.} \cdot 2 + 3^x(0,1) + 1^-(5,2)(Z = 42)$$

$$\doteq 1^o + 7^{1.-6.} \cdot 2 + 4^x + 1^-(Z = 21)$$

$$\doteq 1^o + 7^{1.-6.} \cdot 2 + 5^{1.-2.}(Z = 42)$$

$$\doteq 1^o + 8^{1.-10.} + 3^x(0,3) + 1^-(11,2)(Z = 40)$$

$$\doteq 1^o + 8^{1.-10.} + 4^x + 3^x(0,2) + 1^-(7,1)(Z = 30)$$

$$\doteq 1^o + 8^{1.-10.} + 4^x \cdot 2 + 3^x(0,1) + 1^-(3,0)(Z = 20)$$

$$\doteq 1^o + 8^{1.-10.} + 5^{1.-2.} + 3^x(0,2) + 1^-(6,0)(Z = 60)$$

$$\doteq 1^o + 8^{1.-10.} + 5^{1.-2.} + 4^x + 1^- \cdot 2(Z = 20)$$

$$\doteq 1^o + 8^{1.-10.} + 5^{1.-2.} \cdot 2 + 1^-(Z = 30)$$

$$\doteq 1^o + 8^{1.-10.} + 6^{1.-3.} + 3^x(0,1) + 1^-(5,2)(Z = 60)$$

$$\doteq 1^o + 8^{1.-10.} + 6^{1.-3.} + 4^x + 1^-(Z = 30)$$

$$\doteq 1^o + 8^{1.-10.} + 6^{1.-3.} + 5^{1.-2.}(Z = 60)$$

$$\doteq 1^o + 8^{1.-10.} + 7^{1.-6.} + 3^x(0,1) + 1^-(4,1)(Z = 120)$$

$$\doteq 1^o + 8^{1.-10.} + 7^{1.-6.} + 4^x(Z = 60)$$

$$\doteq 1^o + 8^{1.-10.} \cdot 2 + 3^x(0,1) + 1^-(3,0)(Z = 110)$$

$$\doteq 1^o + 9^{1.-19.} + 3^x(0,3) + 1^-(10,1)(Z = 76)$$

$$\doteq 1^o + 9^{1.-19.} + 4^x + 3^x(0,2) + 1^-(6,0)(Z = 57)$$

$$\doteq 1^o + 9^{1.-19.} + 4^x \cdot 2 + 1^- \cdot 2(Z = 19)$$

$$\doteq 1^o + 9^{1.-19.} + 5^{1.-2.} + 3^x(0,1) + 1^-(5,2)(Z = 76)$$

$$\doteq 1^o + 9^{1.-19.} + 5^{1.-2.} + 4^x + 1^-(Z = 38)$$

$$\doteq 1^o + 9^{1.-19.} + 5^{1.-2.} \cdot 2(Z = 57)$$

$$\doteq 1^o + 9^{1.-19.} + 6^{1.-3.} + 3^x(0,1) + 1^-(4,1)(Z = 114)$$

$$\doteq 1^o + 9^{1.-19.} + 6^{1.-3.} + 4^x(Z = 57)$$

$$\doteq 1^o + 9^{1.-19.} + 7^{1.-6.} + 3^x(0,1) + 1^-(3,0)(Z = 228)$$

$$\doteq 1^o + 9^{1.-19.} + 8^{1.-10.} + 1^- \cdot 2(Z = 190)$$

$$\doteq 1^o + 9^{1.-19.} \cdot 2 + 1^-(Z = 190)$$

Thus

$$h_{20} = 7 + 6 + 4 + 3 + 2 + 10 + 8 + 6 + 2 + 12 +$$
$$6 + 3 + 8 + 4 + 15 + 12 + 6 + 3 + 18 + 12 + 6 +$$
$$18 + 18 + 12 + 12 + 10 + 30 + 18 + 12 + 6 + 36 +$$
$$24 + 18 + 54 + 18 + 36 + 36 + 42 + 21 + 42 + 40 +$$
$$30 + 20 + 60 + 20 + 30 + 60 + 30 + 60 + 120 +$$
$$60 + 110 + 76 + 57 + 19 + 76 + 38 + 57 + 114 +$$
$$57 + 228 + 190 + 190 + 630 = 2988.$$

So

$$h_{20} = 2988 \, .$$

## 1.2.3.8 Enumeration formula for rooted trees

Due to the link between trees and partitions, the following formula can be constructed:

$$r_n = C(P(n-1)) \tag{14}$$

, where C is an entity that takes in a list P of all partitions of the integer n − 1 ($n \geq 2$, with $r_1 = 1$), transforms the partitions into products of binomial coefficients in the same way that the formula (8) does, and then sums up the products and outputs the sum as the quantity for rooted trees with tree size n.

Example:

$$P(6) \equiv 1 \cdot 6 = 2 + 1 \cdot 4 = 2 \cdot 2 + 1 \cdot 2 = 2 \cdot 3 = 3 + 1 \cdot 3 = 3 + 2 + 1 = 3 \cdot 2 = 4 + 1 \cdot 2 = 4 + 2 = 5 + 1 = 6$$

$$C(P(6)) = \binom{r_1 + 6 - 1}{6} + \binom{r_2 + 1 - 1}{1}\binom{r_1 + 4 - 1}{4}$$
$$+ \binom{r_2 + 2 - 1}{2}\binom{r_1 + 2 - 1}{2}$$
$$+ \binom{r_2 + 3 - 1}{3}$$
$$+ \binom{r_3 + 1 - 1}{1}\binom{r_1 + 3 - 1}{3}$$
$$+ \binom{r_3 + 1 - 1}{1}\binom{r_2 + 1 - 1}{1}\binom{r_1 + 1 - 1}{1}$$
$$+ \binom{r_3 + 2 - 1}{2}$$
$$+ \binom{r_4 + 1 - 1}{1}\binom{r_1 + 2 - 1}{2}$$
$$+ \binom{r_4 + 1 - 1}{1}\binom{r_2 + 1 - 1}{1}$$
$$+ \binom{r_5 + 1 - 1}{1}\binom{r_1 + 1 - 1}{1}$$
$$+ \binom{r_6 + 1 - 1}{1}$$

$$C(P(6)) = \binom{r_1 + 5}{6} + \binom{r_2}{1}\binom{r_1 + 3}{4}$$
$$+ \binom{r_2 + 1}{2}\binom{r_1 + 1}{2} + \binom{r_2 + 2}{3}$$
$$+ \binom{r_3}{1}\binom{r_1 + 2}{3} + \binom{r_3}{1}\binom{r_2}{1}\binom{r_1}{1}$$
$$+ \binom{r_3 + 1}{2} + \binom{r_4}{1}\binom{r_1 + 1}{2} + \binom{r_4}{1}\binom{r_2}{1}$$
$$+ \binom{r_5}{1}\binom{r_1}{1} + \binom{r_6}{1}$$

$$C(P(6)) = \binom{6}{6} + \binom{1}{1}\binom{4}{4} + \binom{2}{2}\binom{2}{2} + \binom{3}{3} + \binom{2}{1}\binom{3}{3}$$
$$+ \binom{2}{1}\binom{1}{1}\binom{1}{1} + \binom{3}{2} + \binom{4}{1}\binom{2}{2} + \binom{4}{1}\binom{1}{1}$$
$$+ \binom{9}{1}\binom{1}{1} + \binom{20}{1}$$

$$C(P(6)) = 1 + 1 \cdot 1 + 1 \cdot 1 + 1 + 2 \cdot 1 + 2 \cdot 1 \cdot 1 + 3 + 4 \cdot 1 + 4 \cdot 1 + 9 \cdot 1 + 20$$

$$C(P(6)) = 1 + 1 + 1 + 1 + 2 + 2 + 3 + 4 + 4 + 9 + 20 = 48$$

$$C(P(6)) = 48$$

$$r_7 = 48.$$

# 1.3 Treespeak

Treespeak is an idea of a world where treespeak users, also known as gifted users, use a verbal version of vertex edge algebra to create and destroy. This verbal version, combined with the order defined by the tree generation algorithm, forms treespeak, which gives a possibility to manipulate the graph field that surrounds the user. The manipulation especially includes the trees, and these together with graphs work as the framework for the structures of chemical elements, molecules, atoms, and trajectories in physics. These trees and graphs also have a connection to data structures of program languages, so hacking / programming is possible, in addition to chemistry and physics. Mathematics is however the core of this all, and it can be used to improve and modify one's abilities in treespeak.

The ability of a gifted user to use graphs and trees is measured as a number called tree size metric, or tsm. The higher the tsm, the larger the trees one can use and thus the more complicated and possibly deadly the spells can be. A battle or – tree generation process – between two treespeak users usually culminates to one of the users as winning the battle, and usually this user has a higher tsm. According to estimates, the tsm of the most talented users is over one hundred, but according to the rules of math the tsm should not have any upper bound. The mind of a human may be limited, but what about an entity higher than human? Maybe one theorist knows...

You can check the mathematical tool / game Treespeak in my website www.matladpi.com. Or, alternatively, go to the website https://www.simmer.io/@matladpi/treespeak for the same tool / game with a description. Also, if you are interested, feel free to join the treespeak subreddit r/mathematicaltreespeak in the website www.reddit.com.

# 2.Controversial math

During the years 2014 – 2020 I have researched a couple of controversial topics, such as the 'odd zero' of a modified Riemann zeta function, a 'series deduction method' regarding my dimension theory and the derivative of a discrete binomial coefficient function.

## 2.1 The odd zero of a modified Riemann zeta function

The Riemann zeta function $\zeta(s): \mathbb{C} \setminus \{1\} \to \mathbb{C}$, is defined in the following way:

$$\zeta(s) = \sum_{n=1}^{\infty} n^{-s}, \text{ if } Re(s) > 1 \quad (15)$$

Let us do the following modification to this function:

$$\zeta(s) = \sum_{n=1}^{\infty} n^{-s} = 1^{-s} + 2^{-s} + 3^{-s} + \cdots$$

$$= 1^{-s} + (2^{-s} + 3^{-s} + \cdots)$$

$$\overset{m}{=} 1 + (2^{-s} + 3^{-s} + \cdots) = 1 + \sum_{n=2}^{\infty} n^{-s}$$

$$= g(s)$$

, where m above the equality sign means the modification $1^{-s} = 1$.

Let the parameter s for the function $g(s)$ be defined in the following way:

$Re(s) > 1$.

Let us now show that using 'controversial math' we can find the 'odd zero' $s_0$ for the function $g(s)$, with a property that $Re(s_0) = q_0 > 1$.

$$g(s) = 0$$

$$1 + \sum_{n=2}^{\infty} n^{-s} = 0$$

$$\sum_{n=2}^{\infty} n^{-s} = -1$$

Let us now make a notion $s = \sigma + it$, and we get

$$\sum_{n=2}^{\infty} n^{-(\sigma+it)} = -1$$

$$\sum_{n=2}^{\infty} n^{-\sigma-it} = -1$$

$$\sum_{n=2}^{\infty} n^{-\sigma} n^{-it} = -1$$

Let us now set the case in a way, that $n^{-it} = -1$ for every $n = 2, 3, \dots$ . So now we find t:

$$n^{-it} = -1$$

$$n^{-it} = e^{i\pi + i2\pi k}$$

$$n^{-it} = e^{i\pi(1 + 2k)}$$

$$-it \ln(n) = i\pi(1 + 2k)$$

$$t = \frac{i\pi(1 + 2k)}{-i \ln(n)} = -\frac{\pi(1 + 2k)}{\ln(n)}, \ k \in \mathbb{Z}$$

Thus

$$\sum_{n=2}^{\infty} n^{-\sigma} n^{-it} = -1$$

$$\sum_{n=2}^{\infty} n^{-\sigma} (-1) = -1$$

$(-1) \sum_{n=2}^{\infty} n^{-\sigma} = -1$

$-\sum_{n=2}^{\infty} n^{-\sigma} = -1$

$\sum_{n=2}^{\infty} n^{-\sigma} = 1$

$\sum_{n=2}^{\infty} (n^{-\sigma}) - 1 = 0$

$f(\sigma) = \sum_{n=2}^{\infty} (n^{-\sigma}) - 1$

$f(1{,}5) = \sum_{n=2}^{\infty} (n^{-1{,}5}) - 1 = 0{,}6123 \ldots \approx 0{,}612 > 0$

$f(2) = \sum_{n=2}^{\infty} (n^{-2}) - 1 = \frac{\pi^2}{6} - 2 = -0{,}3550 \ldots \approx$
$-0{,}355 < 0$

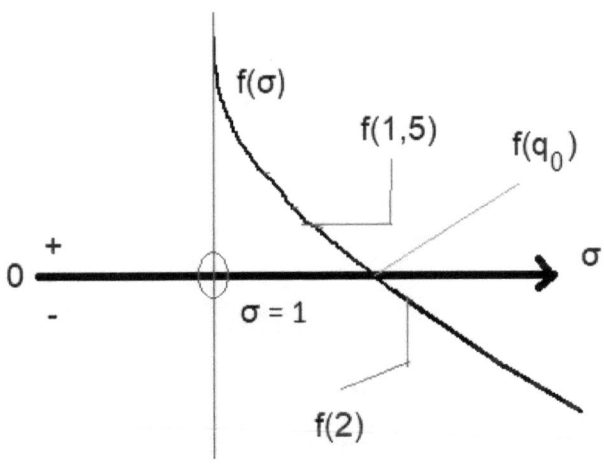

109

Intermediate value theorem: The zero of the continuous and real valued function $f(\sigma)$ $(\sigma > 1)$ is in the area $]\,1{,}5;\,2\,[$, as $f(1{,}5) > 0$ and $f(2) < 0$. Thus the zero has to be the value $q_0$.

The zero $q_0 = 1{,}72864723899818\ldots$ .

Funnily $q_0 \approx e - 1 + 10^{-2}$ and $\displaystyle\lim_{n \to 1} n^{\frac{i\pi}{\ln(n)}} = -1$.

So

$$g(s_0) = 1 + \sum_{n=2}^{\infty} n^{-s_0} = 0$$

, when $s_0 = \sigma + it = q_0 - i\,\dfrac{\pi(1 + 2k)}{\ln(n)}$

, and $i^2 = -1$, $k \in \mathbb{Z}$, $q_0 = 1{,}72864723899818\ldots$ .

# 2.2'Series deduction method' and dimension theory

Dimension theory is a collection of results, which will be deduced after the presentation.

## 2.2.1Dimension theory

Let there be a real number space $\mathbb{R}^D$, (in this theory for the real numbers $\geq 0$) with dimension $D$. Thus, in the dimension $D \in \mathbb{N}\backslash\{0\}$, the following holds:

i.    For an arbitrary transition from a point to another point in dimension $d \in \mathbb{N}\backslash\{0\}$ one needs the dimension $D > d,\ D \in \mathbb{N}\backslash\{0\}$.

ii.   In the space $\mathbb{R}^D$ one can create a restricted space $r^D$ in a way that, for example, $r \in [0, 1]$.

iii.    Any restricted real number space $r^D$ can be divided according to unit grid into small spaces (quantity $n^D$, $n \in \mathbb{N}\backslash\{0\}$), with sizes $n^{-D}$ of the space in question per one small space. These spaces are called $d_1$ -spaces.

iv.    One can create larger d-spaces from $d_1$ -spaces in a way, that one $d_k$ -space contains $k^D$ $d_1$ - spaces.

v.    The quantity of all $d$ -spaces for dimension $D$, when the quantity of $d_1$ -spaces is $n^D$:

$$m_d = \sum_{i=1}^{n} i^D \quad (16)$$

vi.    The quantity of $d_k$ -spaces is $m_{d_k} = (n - k + 1)^D$, $n \geq k$.

Proofs:

i.      One can deduce the following results from the pictures:

1) One can traverse arbitrarily in dimension 1 with dimension 2

2) One can traverse arbitrarily in dimension 2 with dimension 3

Picture 1)

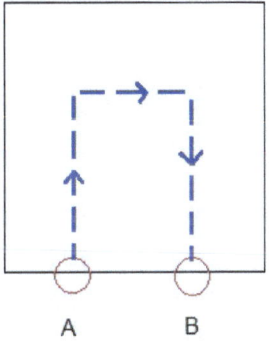

Picture 2)

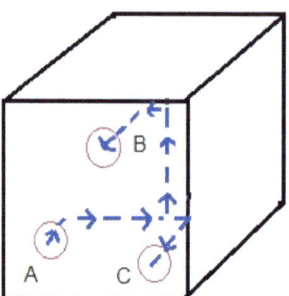

Let us now make a 'series deduction', that for an arbitrary transition from a point to another point in dimension $d \in \mathbb{N}\backslash\{0\}$ one needs the dimension $D > d$, $D \in \mathbb{N}\backslash\{0\}$.

ii.     The result is trivial. No deduction needed.

iii.    The result is trivial. No deduction needed.

iv.  $d_k$ -space contains k $d_1$ -spaces in dimension 1. Thus, one has $k^D$ $d_1$ -spaces in dimension $D$.

v.  Dimension 1 is examined first:

Case L1:

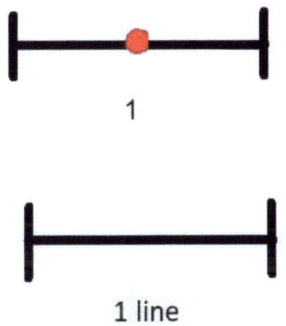

1

1 line

Thus, there is 1 line.

Notice the dot inside the line? It stands for the middle point of the line. In this and the following examples to come the dots always work in the same way and stand for middle points of objects.

Case L2:

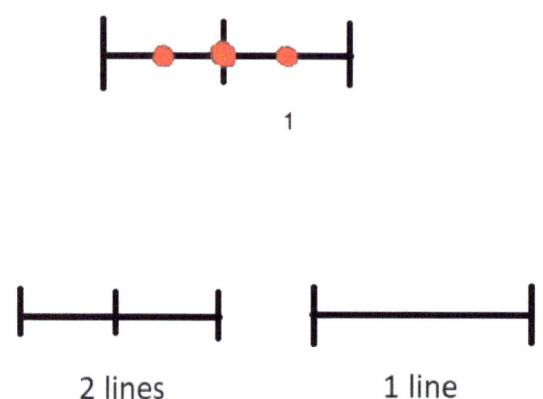

1

2 lines          1 line

Thus, there are $2 + 1 = 3$ lines.

Case L3:

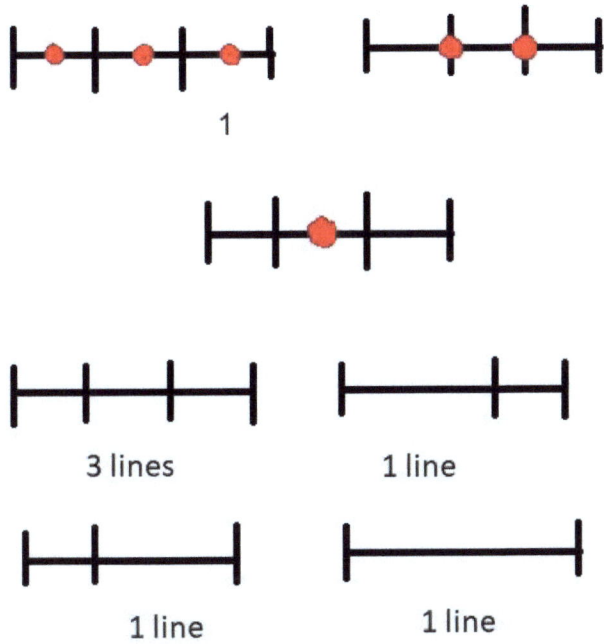

1

3 lines          1 line

1 line            1 line

Thus, there are $3 + 2 + 1 = 6$ lines.

=>

| Case | Number of lines |
|------|-----------------|
| L1   | 1               |
| L2   | 1 + 2           |
| L3   | 1 + 2 + 3       |

An intuitive 'series deduction' is done, that when there are n lines of length 1, one can see

$$1 + 2 + \cdots + (n - 1) + n$$

lines in total.

Dimension 2 is examined now.

Case S1:

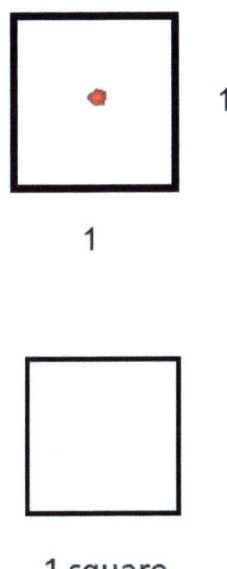

1 square

A single $1 * 1$ square. Thus, we see one square in the picture.

Case S2:

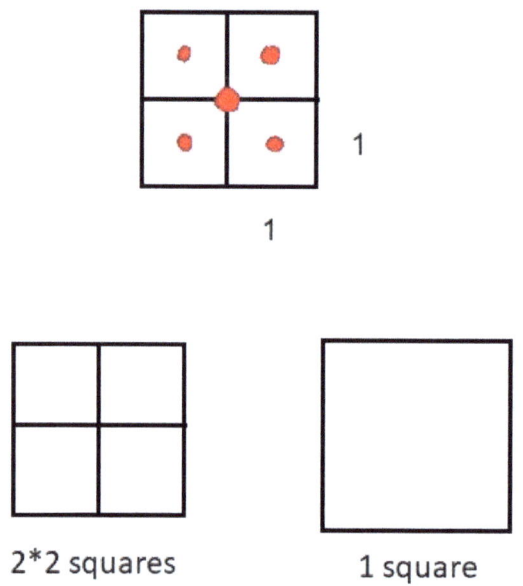

2*2 squares          1 square

There are now $2^2 + 1 = 4 + 1 = 5$ squares.

Case S3:

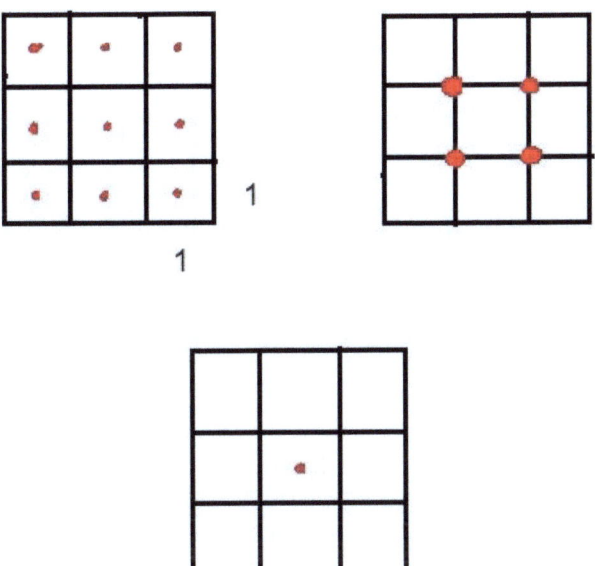

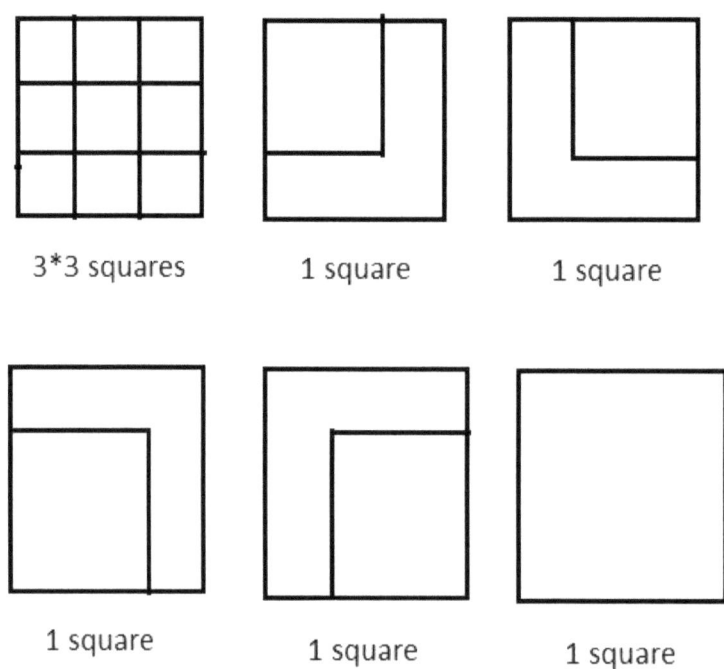

3*3 squares        1 square        1 square

1 square        1 square        1 square

There are now $3^2 + (1 + 1 + 1 + 1) + 1 = 9 + 4 + 1 = 14$ squares.

Notice something? To enlighten the case the sums of the cases are collected:

| Case | Number of squares |
|------|-------------------|
| S1   | 1                 |
| S2   | $1 + 4$           |
| S3   | $1 + 4 + 9$       |

Notice now? The numbers in the sums are square numbers!

In this way:

| Case | Number of squares |
|------|-------------------|
| S1   | $1^2$             |
| S2   | $1^2 + 2^2$       |
| S3   | $1^2 + 2^2 + 3^2$ |

Using intuition, a 'series deduction' can be done that when there are $n^2$ $1 * 1$ squares, one can see

$$1^2 + 2^2 + \cdots + (n-1)^2 + n^2$$

squares in total.

This can also be seen from the dots, which now form the square numbers that correspond to these numbers:

$$1 * 1 + 2 * 2 + \cdots + (n-1) * (n-1) + n * n \,.$$

Dimension 3 is now examined.

Basic cases are examined once again, although the dots are not used now, as the dimension 3 case is more complicated than the dimension 2 or 1.

Case C1:

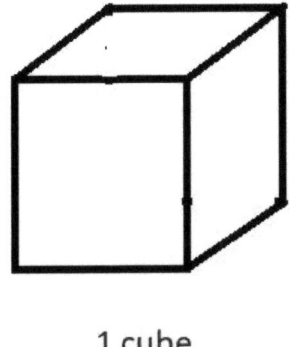

1 cube

There is 1 cube.

Case C2:

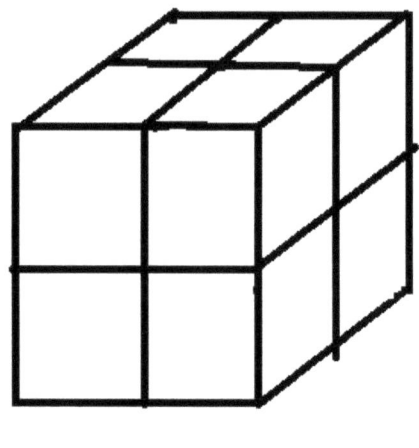

2*2*2 + 1 cubes

There are now $2^3 + 1^3 = 9$ cubes.

Case C3:

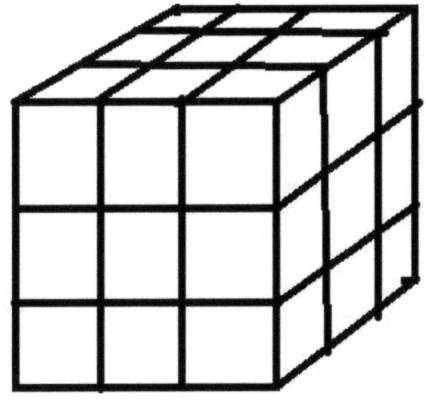

3*3*3 + 2*2*2 + 1 cubes

There are now $3^3 + 2^3 + 1^3 = 36$ cubes.

=>

| Case | Number of cubes |
|------|-----------------|
| C1 | $1^3$ |
| C2 | $1^3 + 2^3$ |
| C3 | $1^3 + 2^3 + 3^3$ |

An intuitive 'series deduction' is now done, that when there are $n^3$ $1 * 1 * 1$ cubes, there are $1^3 + 2^3 + \cdots + n^3$ cubes in total.

The formulas are now collected:

| Dimension | Formula |
|-----------|---------|
| 1 | $\sum_{i=1}^{n} i$ |
| 2 | $\sum_{i=1}^{n} i^2$ |
| 3 | $\sum_{i=1}^{n} i^3$ |

Now the final, intuitive 'series deduction' can be done that, what is the quantity N of all objects, when the object has a dimension D and when the object contains $n^D$ unit objects ($1^D$):

$$N = \sum_{i=1}^{n} i^D$$

This example is now compared to the unit grid and d - spaces and we get

$$m_d = \sum_{i=1}^{n} i^D .$$

This deduction includes the assumption that for objects, dimension 0 corresponds to a point, dimension 1 corresponds to a line, dimension 2 corresponds to a square, dimension 3 corresponds to a cube, dimension 4 corresponds to a hypercube etc.

If the formula is used, then the following formulas are deduced:


| Dimension | Formula |
|---|---|
| 0 | $m_d = \sum_{i=1}^{n} 1$ |
| 1 | $m_d = \sum_{i=1}^{n} i$ |
| 2 | $m_d = \sum_{i=1}^{n} i^2$ |
| 3 | $m_d = \sum_{i=1}^{n} i^3$ |
| 4 | $m_d = \sum_{i=1}^{n} i^4$ |


vi.   For dimension D, n is the highest index in the sum of the number $m_d$, and the calculation goes the way, that

$$(m_{d_k})^{\frac{1}{D}} = n - k + 1 <=> m_{d_k} = (n - k + 1)^D.$$

# 2.3 Derivative of a discrete function

Let there be a binomial coefficient function $B(n) = \binom{b}{n}$, $b \geq n$; $b, n \in \mathbb{N}$.

$$\frac{d}{dn} B(n) = \frac{d}{dn} \binom{b}{n} = \binom{b}{n}(\psi(1 - n + b) - \psi(n + 1))$$

$$\psi(n) = -\gamma + \sum_{k=1}^{n-1} \frac{1}{k}$$

=>

$$\binom{b}{n}(\psi(1 - n + b) - \psi(n + 1))$$

$$= \binom{b}{n}\left(-\gamma + \sum_{k=1}^{b-n} \left(\frac{1}{k}\right) + \gamma - \sum_{k=1}^{n} \left(\frac{1}{k}\right)\right)$$

$$= \binom{b}{n}\left(\sum_{k=1}^{b-n} \left(\frac{1}{k}\right) - \sum_{k=1}^{n} \left(\frac{1}{k}\right)\right) = B'(n)$$

$B'(n) = 0$

$\binom{b}{n} \left( \sum_{k=1}^{b-n} \left( \frac{1}{k} \right) - \sum_{k=1}^{n} \left( \frac{1}{k} \right) \right) = 0$

$\binom{b}{n} > 0$ for every $b \geq n$; $b, n \in \mathbb{N}$

=>

$\left( \sum_{k=1}^{b-n} \left( \frac{1}{k} \right) - \sum_{k=1}^{n} \left( \frac{1}{k} \right) \right) = 0$

$\sum_{k=1}^{b-n} \left( \frac{1}{k} \right) - \sum_{k=1}^{n} \left( \frac{1}{k} \right) = 0$

$\sum_{k=1}^{b-n} \left( \frac{1}{k} \right) = \sum_{k=1}^{n} \left( \frac{1}{k} \right)$

The equation holds, when

$b - n = n <=> 2n = b <=> n = \frac{b}{2}$

The zero of the derivative of the function is $\frac{b}{2} = p$, when $b = 2p, p = 1,2,3, \dots$ .

Let us now show, that $B\left(\frac{b}{2}\right)$ is the maximum value of the function:

The following must hold:

1) $B'\left(\frac{b}{2} - m\right) > 0, 0 < m < \frac{b}{2}, b = 2p$

2) $B'\left(\frac{b}{2} + m\right) < 0, 0 < m < \frac{b}{2}, b = 2p$

1) $B'\left(\frac{b}{2} - m\right) > 0, 0 < m < \frac{b}{2}, b = 2p$

$$\left(\frac{2p}{p-m}\right)\left(\sum_{k=1}^{p+m}\left(\frac{1}{k}\right) - \sum_{k=1}^{p-m}\left(\frac{1}{k}\right)\right) > 0$$

$\left(\frac{2p}{p-m}\right) > 0$ for every $0 < m < p$

=>

$$\left(\sum_{k=1}^{p+m} \left(\frac{1}{k}\right) - \sum_{k=1}^{p-m} \left(\frac{1}{k}\right)\right) > 0$$

$$\sum_{k=1}^{p+m} \left(\frac{1}{k}\right) - \sum_{k=1}^{p-m} \left(\frac{1}{k}\right) > 0$$

$$\left(\sum_{k=p-m+1}^{p+m} \left(\frac{1}{k}\right) + \sum_{k=1}^{p-m} \left(\frac{1}{k}\right)\right) - \sum_{k=1}^{p-m} \left(\frac{1}{k}\right) > 0$$

$$\sum_{k=p-m+1}^{p+m} \left(\frac{1}{k}\right) + \sum_{k=1}^{p-m} \left(\frac{1}{k}\right) - \sum_{k=1}^{p-m} \left(\frac{1}{k}\right) > 0$$

$$\sum_{k=p-m+1}^{p+m} \left(\frac{1}{k}\right) > 0$$

$0 < m < p => k \geq p - m + 1 > m - m + 1 = 1 > 0$

$p + m \geq 1 + m > 1 + 0 = 1$

$p + m > p - m + 1 <=> 2m > 1$. True, as $2m \geq 2 \cdot 1 = 2$.

=> the statement holds, thus

1) $B'\left(\frac{b}{2} - m\right) > 0, 0 < m < \frac{b}{2}, b = 2p$

2) $B'\left(\frac{b}{2}+m\right) < 0, 0 < m < \frac{b}{2}, b = 2p$

$$\binom{2p}{p+m}\left(\sum_{k=1}^{p-m}\left(\frac{1}{k}\right) - \sum_{k=1}^{p+m}\left(\frac{1}{k}\right)\right) < 0$$

$$\binom{2p}{p+m} > 0 \text{ for every } 0 < m < p$$

$$\left(\sum_{k=1}^{p-m}\left(\frac{1}{k}\right) - \sum_{k=1}^{p+m}\left(\frac{1}{k}\right)\right) < 0$$

$$\sum_{k=1}^{p-m}\left(\frac{1}{k}\right) - \left(\sum_{k=p-m+1}^{p+m}\left(\frac{1}{k}\right) + \sum_{k=1}^{p-m}\left(\frac{1}{k}\right)\right) < 0$$

$$\sum_{k=1}^{p-m}\left(\frac{1}{k}\right) - \sum_{k=1}^{p-m}\left(\frac{1}{k}\right) - \sum_{k=p-m+1}^{p+m}\left(\frac{1}{k}\right) < 0$$

$$-\sum_{k=p-m+1}^{p+m}\left(\frac{1}{k}\right) < 0$$

$$\sum_{k=p-m+1}^{p+m}\left(\frac{1}{k}\right) > 0$$

$0 < m < p \Rightarrow k \geq p - m + 1 > m - m + 1 = 1 > 0$

$p + m \geq 1 + m > 1 + 0 = 1$

$p + m > p - m + 1 \Leftrightarrow 2m > 1$. True, as $2m \geq 2 \cdot 1 = 2$.

=> the statement holds, thus

2) $B'\left(\frac{b}{2} + m\right) < 0, 0 < m < \frac{b}{2}$ , $b = 2p$

Both cases 1) and 2) have been proven, thus the point $n = \frac{b}{2}$ gives the maximum value for the binomial coefficient function $B(n) = \binom{b}{n}$.

# 3.Good old constant

## 3.1Pi

$\pi$ is the mathematical constant pi,

$$\pi = 3{,}14159\ldots$$

, for which the following limit can be deduced:

$$\pi = \lim_{n \to \infty} \left( n \sin \frac{180°}{n} \right) \quad (17)$$

This result can be deduced from a regular 'n-gon' that is inside a circle.

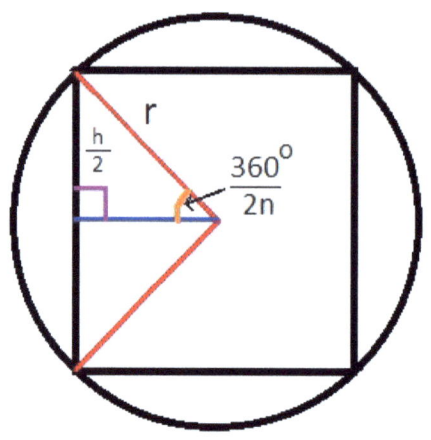

$$\sin\frac{360°}{2n} = \frac{h}{2r}$$

$$\sin\frac{180°}{n} = \frac{h}{2r}$$

$$h = 2r\sin\frac{180°}{n}$$

$$\lim_{n \to \infty} C_{n-gon} = C_{circle}$$

$$\lim_{n \to \infty} (nh) = 2\pi r$$

$$\lim_{n \to \infty} \left(2nr\sin\frac{180°}{n}\right) = 2\pi r$$

$$2\pi r = \lim_{n \to \infty} \left(2nr\sin\frac{180°}{n}\right)$$

$$\pi = \lim_{n \to \infty} \left( n \sin \frac{180°}{n} \right)$$

With the information $360° = 2\pi \ (rad) <=> 180° = \pi \ (rad)$ it can be deduced, that

$$\lim_{n \to 0+} \left( \frac{\sin n}{n} \right) = \lim_{n \to 0-} \left( \frac{\sin n}{n} \right) = \lim_{n \to 0} \left( \frac{\sin n}{n} \right) =$$
$$\lim_{n \to 0} \left( \frac{n}{n} \right) = \lim_{n \to 0} (1) = 1$$

This result is now deduced:

$$\lim_{n \to \infty} \left( n \sin \frac{180°}{n} \right) = \pi$$

$$\lim_{n \to \infty} \left( n \sin \frac{\pi}{n} \right) = \pi$$

$$\lim_{n \to \infty} \left( \sin \left( \frac{\pi}{n} \right) \right) = \lim_{n \to \infty} \left( \frac{\pi}{n} \right)$$

$$\lim_{m \to 0-} (\sin(\pi m)) = \lim_{m \to 0-} (\pi m)$$

$$\lim_{m \to 0+} (\sin(\pi m)) = \lim_{m \to 0+} (\pi m)$$

$$\lim_{m \to 0} (\sin(\pi m)) = \lim_{m \to 0} (\pi m)$$

$$\lim_{n \to 0} \sin(n) = \lim_{n \to 0} (n) = 0$$

=>

$$\lim_{n \to 0} \left( \frac{\sin(n)}{n} \right) = \lim_{n \to 0} \left( \frac{n}{n} \right) = \lim_{n \to 0} (1) = 1$$

So

$$\lim_{n \to 0} \left( \frac{\sin(n)}{n} \right) = 1 \,.$$

## 3.2 Euler's number e

$e$ is the Euler's number,

$$e = 2{,}71828 \dots$$

, for which the following series can be deduced:

$$e = \sum_{k=0}^{\infty} \frac{1}{k!} \quad (18)$$

Deduction of the result:

$$e^t = \lim_{n \to \infty} \left(1 + \frac{t}{n}\right)^n$$

$$= \lim_{n \to \infty} \Sigma_{k=0}^n \left(\binom{n}{k} 1^{n-k} \left(\frac{t}{n}\right)^k\right)$$

$$= \lim_{n \to \infty} \Sigma_{k=0}^n \left(\binom{n}{k} n^{-k} (t)^k\right)$$

$$= \lim_{n \to \infty} \left(\binom{n}{0}\left(\frac{t}{n}\right)^0 + \binom{n}{1}\left(\frac{t}{n}\right)^1 + \binom{n}{2}\left(\frac{t}{n}\right)^2 + \binom{n}{3}\left(\frac{t}{n}\right)^3 + \cdots\right)$$

$$= \lim_{n \to \infty} \left(1 + \frac{n}{n}t + \frac{n(n-1)}{2n^2}t^2 + \frac{n(n-1)(n-2)}{6n^3}t^3 + \cdots\right)$$

$$= \lim_{n \to \infty} \left(1 + \left(\frac{t}{1}\right)\left(\frac{n}{n}\right) + \left(\frac{t^2}{2}\right)\left(\frac{n}{n}\right)\left(\frac{n-1}{n}\right) + \left(\frac{t^3}{6}\right)\left(\frac{n}{n}\right)\left(\frac{n-1}{n}\right)\left(\frac{n-2}{n}\right) + \cdots\right)$$

$$= \frac{t^0}{0!} + \frac{t^1}{1!} * 1 + \frac{t^2}{2!} * 1 * 1 + \frac{t^3}{3!} * 1 * 1 * 1 + \cdots$$

$$= \frac{t^0}{0!} + \frac{t^1}{1!} + \frac{t^2}{2!} + \frac{t^3}{3!} + \cdots$$

$$= \sum_{k=0}^{\infty} \frac{t^k}{k!}$$

So

$$e^t = \sum_{k=0}^{\infty} \frac{t^k}{k!}$$

and

$$e^1 = e = \sum_{k=0}^{\infty} \frac{1}{k!}.$$

# 3.3 Euler-Mascheroni constant

For the Euler-Mascheroni constant,

$$\gamma = 0,57721 \dots$$

142

, the following holds:

$$\gamma = \sum_{i=0}^{n-1}\left(\frac{1}{n-i}\right) - \frac{d}{dn}\sum_{i=0}^{n-1}\ln(n-i) \quad (19)$$

$$n! = \prod_{i=0}^{n-1}(n-i) = \exp\left(\sum_{i=0}^{n-1}\ln(n-i)\right)$$

$$\frac{d}{dn}(n!) = \frac{d}{dn}\exp\left(\sum_{i=0}^{n-1}\ln(n-i)\right)$$

$$= \exp\left(\sum_{i=0}^{n-1}\ln(n-i)\right)\frac{d}{dn}\sum_{i=0}^{n-1}\ln(n-i)$$

$$= n!\frac{d}{dn}\sum_{i=0}^{n-1}\ln(n-i)$$

$$\frac{d}{dn}(n!) = n!\left(-\gamma + \sum_{k=1}^{n}\frac{1}{k}\right)$$

$$n!\frac{d}{dn}\sum_{i=0}^{n-1}\ln(n-i) = n!\left(-\gamma + \sum_{k=1}^{n}\frac{1}{k}\right)$$

$$\frac{d}{dn}\sum_{i=0}^{n-1}\ln(n-i) = -\gamma + \sum_{k=1}^{n}\frac{1}{k}$$

$$\frac{d}{dn}\sum_{i=0}^{n-1}\ln(n-i) = -\gamma + \left(\frac{1}{1}+\frac{1}{2}+\frac{1}{3}+\cdots+\frac{1}{n-1}+\frac{1}{n}\right)$$

$$\frac{d}{dn}\sum_{i=0}^{n-1}\ln(n-i) = -\gamma + \left(\frac{1}{n}+\frac{1}{n-1}+\cdots+\frac{1}{3}+\frac{1}{2}+\frac{1}{1}\right)$$

$$\frac{d}{dn}\sum_{i=0}^{n-1}\ln(n-i) = -\gamma + \sum_{i=0}^{n-1}\frac{1}{n-i} \quad (20)$$

$$\gamma = \sum_{i=0}^{n-1}\left(\frac{1}{n-i}\right) - \frac{d}{dn}\sum_{i=0}^{n-1}\ln(n-i)$$

# 4.A theorem in geometry

If the largest possible circle (every circle is identical to others) is put in every internal angle of a regular n-gon, then the quotient between the area between the circles and the area of the whole n-gon is

$$\eta(n) = \frac{1 - \dfrac{n-2}{2n}\pi \tan\dfrac{\pi}{n}}{\left(1 + \tan\dfrac{\pi}{n}\right)^2}$$

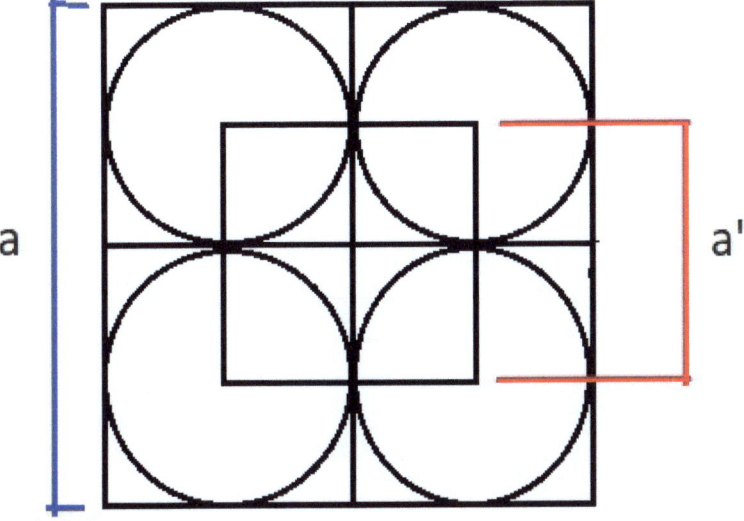

There is another n-gon inside the n-gon, and the vertices of this n-gon are the middle points of the circles.

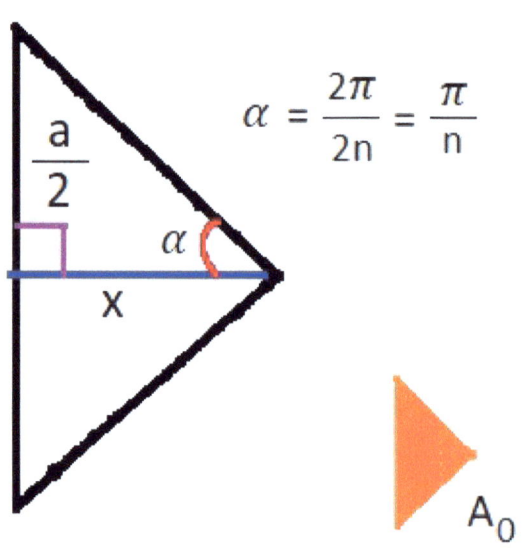

$$\alpha = \frac{2\pi}{2n} = \frac{\pi}{n}$$

$A_0$

$$\tan\frac{\pi}{n} = \frac{a}{2x} \iff x = \frac{a}{2\tan\frac{\pi}{n}}$$

$$A_0 = \frac{ax}{2} = \frac{a^2}{4\tan\frac{\pi}{n}}$$

The area of an a-sided, regular n-gon:

$$A_{an} = nA_0 = \frac{na^2}{4\tan\frac{\pi}{n}} \quad (21)$$

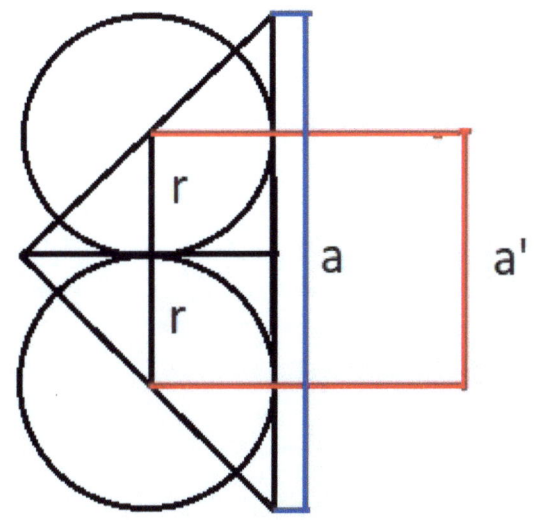

a′ = 2r

Let us come up with a formula for the quotient

$$\frac{r}{a} = q(n)$$

$q_3 \approx 0{,}1830127$

$q_4 \approx 0{,}2500000$

$q_5 \approx 0{,}2895961$

$q_6 \approx 0{,}3169873$ .

Let us deduce that $\dfrac{r}{a} = q(n) = \dfrac{1}{2\left(1+\tan\frac{\pi}{n}\right)} <=> r = $

$\dfrac{a}{2\left(1+\tan\frac{\pi}{n}\right)}$

, and $a' = 2r = \dfrac{a}{1+\tan\frac{\pi}{n}}$ .

The area of an a'-sided, regular n-gon:

$$A_{a'n} = \frac{n(a')^2}{4\tan\frac{\pi}{n}} = \frac{n}{4\tan\frac{\pi}{n}}\frac{a^2}{\left(1+\tan\frac{\pi}{n}\right)^2} = \frac{na^2}{4\tan\frac{\pi}{n}\left(1+\tan\frac{\pi}{n}\right)^2}$$

$$A_{circles} = \frac{n-2}{2}A_{circle} = \frac{n-2}{2}\pi r^2 = \frac{n-2}{2}\pi\frac{a^2}{4\left(1+\tan\frac{\pi}{n}\right)^2}$$

=>

$$\eta(n) = \frac{A_{aln} - A_{circles}}{A_{an}}$$

$$= \left( \frac{na^2}{4\tan\frac{\pi}{n}\left(1+\tan\frac{\pi}{n}\right)^2} - \frac{n-2}{2}\pi\frac{a^2}{4\left(1+\tan\frac{\pi}{n}\right)^2} \right)\left( \frac{na^2}{4\tan\frac{\pi}{n}} \right)^{-1}$$

$$= \left( \frac{na^2}{4\tan\frac{\pi}{n}\left(1+\tan\frac{\pi}{n}\right)^2} - \frac{n-2}{2}\pi\frac{a^2\tan\frac{\pi}{n}}{4\tan\frac{\pi}{n}\left(1+\tan\frac{\pi}{n}\right)^2} \right)\left( \frac{na^2}{4\tan\frac{\pi}{n}} \right)^{-1}$$

$$= \left( \frac{na^2 - \frac{n-2}{2}\pi a^2\tan\frac{\pi}{n}}{4\tan\frac{\pi}{n}\left(1+\tan\frac{\pi}{n}\right)^2} \right)\frac{4\tan\frac{\pi}{n}}{na^2}$$

$$= \frac{1 - \frac{n-2}{2n}\pi\tan\frac{\pi}{n}}{\left(1+\tan\frac{\pi}{n}\right)^2}$$

Values:

$\eta(3) = 0{,}01247\ldots$

$\eta(4) = 0{,}05365\ldots$

$\eta(5) = 0{,}10575\ldots$

$\eta(6) = 0{,}15892\ldots$

$\eta(7) = 0{,}20941\ldots$

$\eta(8) = 0{,}25600\ldots$

$\eta(9) = 0{,}29849\ldots$

$\eta(10) = 0{,}33706\ldots$

$\eta(20) = 0{,}57837\ldots$

$\eta(50) = 0{,}80114\ldots$

$\eta(100) = 0{,}89451\ldots$

$\eta(200) = 0{,}94562\ldots$

$\eta(500) = 0{,}97784\ldots$

$\eta(1000) = 0{,}98885\ldots$

$\lim\limits_{n \to \infty} \eta(n) = 1$ .

# 5.Identity number and the golden ratio

Identity number is about the structure of a tree and is divided into two numbers:

1) The number $\ell_- \in [0, 1]$, $\ell_- \neq \frac{n-3}{n-1}, \frac{n-2}{n-1}$, when $n$ is the number of a tree's vertices and $\ell_-$ is not canceled out

2) The number $\ell_x \in [0,1]$, $\ell_x \neq \frac{1}{n-1}, \frac{2}{n-1}$, when $n$ is the number of a tree's vertices and $\ell_x$ is not canceled out

From these, the rational number (finite tree) $\ell_-$ is about the longitudinal nature of the tree. The larger this number is, the more longitudinal the tree is.

The rational number (finite tree) $\ell_x$, on the other hand, is about the tree's furcate nature. The larger this number is, the more furcate the tree is.

These numbers can also be turned into percentages, although the original value that is not canceled out should be preserved. Namely, this number contains some original information about the tree, for example the number of its edges.

The number $\ell_-$ can be calculated in the following way:

Choose a vertex, which is connected to at least three other vertices (if there is no such vertex, then $\ell_- = 1$) with edges. The counting happens with going outwards from the vertex in question. Count the edges in all the longitudinal parts, in which there are at least two or more edges in a line. In these parts there are no branches between the edges. When you have calculated the total number of edges in these parts, subtract the number of the longitudinal parts from the total number and then divide the result by the total number of the edges in the tree.

The number $\ell_x$ can be calculated in the following way:

Choose a vertex, which is connected to at least three other vertices (if there is no such vertex, then $\ell_x = 0$) with edges. The examination happens outwards from this vertex, and count every branch in vertices, which are connected to at least three other vertices with edges. Thus, from the perspective of the original vertex, there are two branches connected to the vertices, aligned outward. Inside branches there may also be other branches etc. Divide the result by the total number of edges in the tree.

The following holds:

$$\ell_- + \ell_x = 1 \quad (22)$$

Examples:

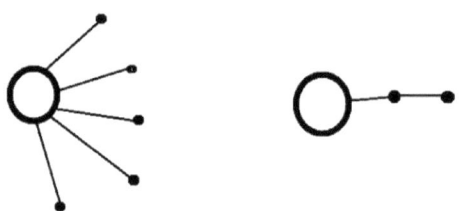

$$\ell_- = \frac{0}{5} = 0 \,, \ell_x = \frac{5}{5} = 1 \quad \ell_- = \frac{2}{2} = 1 \,, \; \ell_x = \frac{0}{2} = 0$$

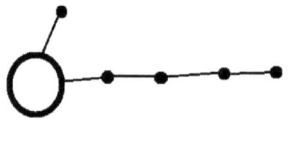 

$$\ell_- = \frac{5}{5} = 1, \ell_x = \frac{0}{5} = 0 \qquad \ell_- = \frac{2}{2} = 1, \; \ell_x = \frac{0}{2} = 0$$

$\ell_-, \ell_x$ not defined

$\ell_-, \ell_x$ not defined

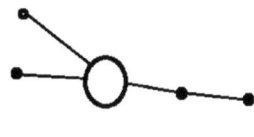

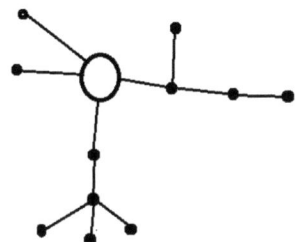

$$\ell_- = \frac{1}{4}, \ell_x = \frac{3}{4}$$

$$\ell_- = \frac{2}{11}, \ell_x = \frac{9}{11}$$

$\varphi^{-3}$ is the number one divided by the golden ratio

$$\varphi = \frac{1+\sqrt{5}}{2} = 1{,}61803 \ldots$$

, to the third power. The number $\varphi^{-3}$ is associated with the number $\ell_-$ and Collatz conjecture:

$\ell_- = \varphi^{-3} = 0{,}23606 \ldots$ for the infinite $3n + 1/2^{-1}$ - diagram of the Collatz conjecture.

Collatz conjecture:

Let there be the following iteration algorithm for positive integers:

$n \to 3n + 1$ , when $n$ is odd

$n \to \frac{n}{2}$ , when $n$ is even.

With any n, the final result will be one.

The diagram:

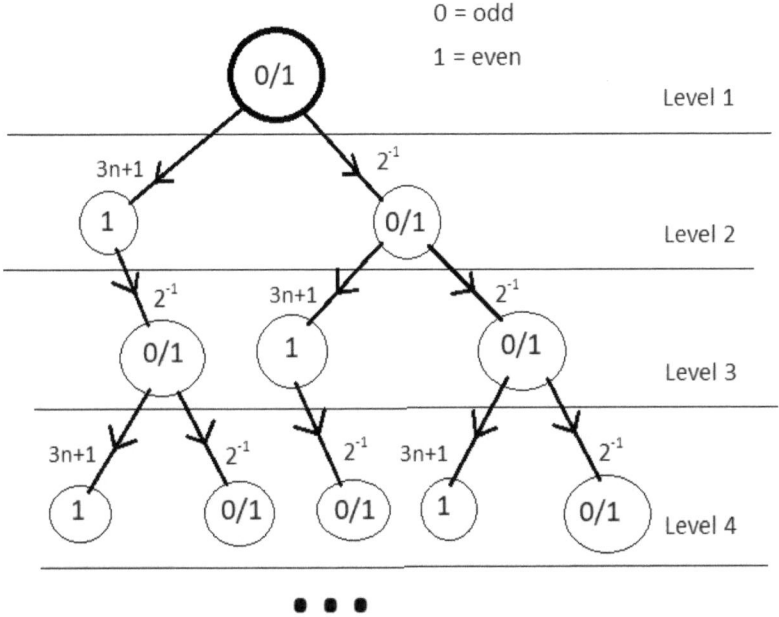

Let us mark the $n$. Fibonacci number with $F_n$ , when

$$F_1 = F_2 = 1$$

and

$$F_n = F_{n-1} + F_{n-2}, n \geq 3.$$

The number $\ell_-$ changes according to the diagram's levels in the following way:

| Level | $\ell_-$ |
|---|---|
| 1 | not defined |
| 2 | $\dfrac{2}{2}$ |
| 3 | $\dfrac{(1)+1}{(2)+3} = \dfrac{2}{5}$ |
| 4 | $\dfrac{(1)+1+1}{(2)+3+5} = \dfrac{3}{10}$ |
| 5 | $\dfrac{(1)+1+1+2}{(2)+3+5+8} = \dfrac{5}{18}$ |
| 6 | $\dfrac{(1)+1+1+2+3}{(2)+3+5+8+13} = \dfrac{8}{31}$ |
| 7 | $\dfrac{(1)+F_1+F_2+F_3+F_4+F_5}{(2)+F_4+F_5+F_6+F_7+F_8} = \dfrac{13}{52}$ |
| n | $\dfrac{(1)+\sum_{i=1}^{n-2} F_i}{(2)+\sum_{i=4}^{n+1} F_i}$ |

$$\frac{(1) + \sum_{i=1}^{n-2} F_i}{(2) + \sum_{i=4}^{n+1} F_i} = \frac{1 + \sum_{i=1}^{n-2} F_i}{2 + \sum_{i=4}^{n+1} F_i}$$

$$= \frac{1 + \sum_{i=1}^{n-2} F_i}{2 + (4 - 4) + \sum_{i=4}^{n+1} F_i}$$

$$= \frac{1 + \sum_{i=1}^{n-2} F_i}{2 + 4 - 4 + \sum_{i=4}^{n+1} F_i}$$

$$= \frac{1 + \sum_{i=1}^{n-2} F_i}{(2 - 4) + \left(\sum_{i=4}^{n+1} F_i + 4\right)}$$

$$= \frac{(1) + \sum_{i=1}^{n-2} F_i}{(-2) + \left(\sum_{i=4}^{n+1} F_i + 1 + 1 + 2\right)}$$

$$= \frac{(1) + \sum_{i=1}^{n-2} F_i}{(-2) + \left(\sum_{i=4}^{n+1} F_i + F_1 + F_2 + F_3\right)}$$

$$= \frac{(1) + \sum_{i=1}^{n-2} F_i}{(-2) + \left(\sum_{i=1}^{n+1} F_i\right)}$$

$$= \frac{(1) + \sum_{i=1}^{n-2} F_i}{(-2) + \sum_{i=1}^{n+1} F_i}$$

The following holds:

$$\sum_{i=1}^{n} F_i = F_{n+2} - 1$$

=>

$$\ell_- = \frac{(1) + \sum_{i=1}^{n-2} F_i}{(-2) + \sum_{i=1}^{n+1} F_i}$$

$$= \frac{(1) + \left(\sum_{i=1}^{n-2} F_i\right)}{(-2) + \left(\sum_{i=1}^{n+1} F_i\right)}$$

$$= \frac{(1) + (F_{(n-2)+2} - 1)}{(-2) + (F_{(n+1)+2} - 1)}$$

$$= \frac{F_n}{F_{n+3} - 3}$$

, n ≥ 3

$\ell_- = \dfrac{F_n}{F_{n+3}-3}$ for a finite $3n+1/2^{-1}$ -diagram with n levels (n ≥ 3), when the diagram is related to Collatz conjecture.

=>

$$\lim_{n \to \infty} \frac{F_n}{F_{n+3}-3} = \lim_{n \to \infty} \frac{F_n}{F_{n+3}}$$

$$= \lim_{n \to \infty} \frac{F_n F_{n+1} F_{n+2}}{F_{n+3} F_{n+1} F_{n+2}}$$

$$= \lim_{n \to \infty} \frac{F_n F_{n+1} F_{n+2}}{F_{n+1} F_{n+2} F_{n+3}}$$

$$= \lim_{n \to \infty} \left(\frac{F_n}{F_{n+1}}\right)\left(\frac{F_{n+1}}{F_{n+2}}\right)\left(\frac{F_{n+2}}{F_{n+3}}\right)$$

$$= \lim_{n \to \infty} \left(\frac{F_{n+1}}{F_n}\right)^{-1}\left(\frac{F_{n+2}}{F_{n+1}}\right)^{-1}\left(\frac{F_{n+3}}{F_{n+2}}\right)^{-1}$$

$$= (\varphi)^{-1}(\varphi)^{-1}(\varphi)^{-1} = \varphi^{-1-1-1} = \varphi^{-3}$$

$\ell_- = \varphi^{-3} = 0{,}23606\ldots$ for the infinite $3n+1/2^{-1}$ - diagram of the Collatz conjecture.

## A fun fact

$n$ , $n$ is an odd, positive integer

$3n + 1$

$(3n + 1)2^{-k_1} = 3n * 2^{-k_1} + 2^{-k_1}$

$3(3n * 2^{-k_1} + 2^{-k_1}) + 1 = 3^2 n * 2^{-k_1} + 3 * 2^{-k_1} + 1$

$(3^2 n * 2^{-k_1} + 3 * 2^{-k_1} + 1)2^{-k_2} = 3^2 n * 2^{-k_1 - k_2} +$
$3 * 2^{-k_1 - k_2} + 2^{-k_2}$

$3(3^2 n * 2^{-k_1 - k_2} + 3 * 2^{-k_1 - k_2} + 2^{-k_2}) + 1$

$= 3^3 n * 2^{-k_1 - k_2} + 3^2 * 2^{-k_1 - k_2} + 3 * 2^{-k_2} + 1$

$(3^3 n * 2^{-k_1 - k_2} + 3^2 * 2^{-k_1 - k_2} + 3 * 2^{-k_2} + 1)2^{-k_3}$

$= 3^3 n * 2^{-k_1 - k_2 - k_3} + 3^2 * 2^{-k_1 - k_2 - k_3} + 3 *$
$2^{-k_2 - k_3} + 2^{-k_3}$

$$3^p n * 2^{-k_1-k_2-\cdots-k_p} + 3^{p-1} * 2^{-k_1-k_2-\cdots-k_p} + 3^{p-2} *$$
$$2^{-k_2-\cdots-k_p} + \cdots + 3 * 2^{-k_{p-1}-k_p} + 3^0 * 2^{-k_p} = 1$$

$$(3n+1)3^{p-1} \prod_{i=1}^{p} 2^{-k_i} + 3^{p-2} \prod_{i=2}^{p} 2^{-k_i} +$$
$$3^{p-3} \prod_{i=3}^{p} 2^{-k_i} + \cdots + 2^{-k_p} = 1$$

$$(3n+1)3^{p-1} \prod_{i=1}^{p} 2^{-k_i} + \sum_{j=2}^{p} \left( 3^{p-j} \prod_{i=j}^{p} 2^{-k_i} \right) = 1$$

$, p, k_i \in \mathbb{N}; \ p \geq 3, k_i \geq 1 .$

$m$ , $m$ is an even, positive integer

$$2^{-h_1} m$$

$$3(2^{-h_1}m) + 1 = 3 * 2^{-h_1}m + 1$$

$$(3 * 2^{-h_1}m + 1)2^{-h_2} = 3 * 2^{-h_1-h_2}m + 2^{-h_2}$$

$3(3 * 2^{-h_1-h_2}m + 2^{-h_2}) + 1 = 3^2 * 2^{-h_1-h_2}m + 3 * 2^{-h_2} + 1$

$(3^2 * 2^{-h_1-h_2}m + 3 * 2^{-h_2} + 1)2^{-h_3}$

$= 3^2 * 2^{-h_1-h_2-h_3}m + 3 * 2^{-h_2-h_3} + 2^{-h_3}$

$3^q m * 2^{-h_1-h_2-h_3-\cdots-h_q-h_{q+1}} + 3^{q-1} *$
$2^{-h_2-h_3-\cdots-h_q-h_{q+1}} + \cdots + 3^0 * 2^{-h_{q+1}} = 1$

$3^q m \prod_{i=1}^{q+1} 2^{-h_i} + 3^{q-1} \prod_{i=2}^{q+1} 2^{-h_i} + \cdots +$
$3^{q-q} \prod_{i=q+1}^{q+1} 2^{-h_i} = 1$

$$3^q m \prod_{i=1}^{q+1} 2^{-h_i} + \sum_{j=1}^{q} \left( 3^{q-j} \prod_{i=j+1}^{q+1} 2^{-h_i} \right) = 1$$

$, q, h_i \in \mathbb{N}, q \geq 1, h_i \geq 1.$

# 6. A theorem for number one

The number one can be expressed as a product (arbitrary length, but at least two numbers) of irrational numbers, which are greater than zero and smaller than three.

Any nonzero, positive integer $k \neq p^2 - 1, p \in \mathbb{N}, p \geq 2$; can be expressed as a product of irrational numbers $r_i$, where $0 < r_i \leq \sqrt{k+1} + 1$.

$$1 = 2 - 1 = 2^{\left(\frac{1}{2}\right)*2} - 1^2$$

$$= \left(2^{\frac{1}{2}}\right)^2 - 1^2$$

$$= \left(2^{\frac{1}{2}} - 1\right)\left(2^{\frac{1}{2}} + 1\right)$$

$$= \left(2^{\left(\frac{1}{4}\right)*2} - 1^2\right)\left(2^{\frac{1}{2}} + 1\right)$$

$$= \left(2^{\frac{1}{4}} - 1\right)\left(2^{\frac{1}{4}} + 1\right)\left(2^{\frac{1}{2}} + 1\right)$$

$$= \left(2^{\left(\frac{1}{8}\right)*2} - 1^2\right)\left(2^{\frac{1}{4}} + 1\right)\left(2^{\frac{1}{2}} + 1\right)$$

$$= \left(2^{\frac{1}{8}} - 1\right)\left(2^{\frac{1}{8}} + 1\right)\left(2^{\frac{1}{4}} + 1\right)\left(2^{\frac{1}{2}} + 1\right)$$

...

$$1 = \left(2^{2^{-n}} - 1\right)\prod_{i=1}^{n}\left(2^{2^{-i}} + 1\right)$$

, for every $n \in \mathbb{N}$, $n \geq 1$

$$k = \left((k+1)^{2^{-n}} - 1\right)\prod_{i=1}^{n}\left((k+1)^{2^{-i}} + 1\right)$$

, for every $k, n \in \mathbb{N}$; $k, n \geq 1, k \neq p^2 - 1, p \in \mathbb{N}, p \geq 2$.